# The Essential Guide
# to Managing a Government Project

*What Every Project Manager Should Know*

By Michael Lisagor

*The Government FreeLance Exchange (GovFlex.com)*

*Government Contracting Knowledge Series*

Published by The Government FreeLance Exchange (GovFlex.com)

Printed in the United States of America

10 9 8 7 6 5 4 3 2 1

Library of Congress Cataloging-in-Publication Data

The Essential Guide to Managing a Government Project

ISBN: 9781451568738

Nonfiction > Technology & Engineering > Project Management

Nonfiction > Computers > Information Technology

# Table of Contents

3

# Foreword

Federal as well as State & Local government projects are too often plagued by cost overruns, schedule delays and technical deficiencies. One of many contributing factors is that project managers (PMs) continue to make the same mistakes. And, while Program Management Institute (PMI) training is invaluable, once a PM gets thrown into their first project, most of this knowledge falls by the wayside. Plus some of the fundamentals don't directly translate to the government arena. The purpose of this Guide is to bridge the gap between formal training and hands-on experience by sharing What to Do and What Not to Do.

*The Essential Guide to Managing a Government Project* covers what an industry or government PM needs to know to successfully complete a government development or services contract. I've tried to use an easy-to-read style of writing to make project management simple and readable so PMs don't just permanently banish this Guide to their bookshelves.

I've done my best to keep the contents consistent with PMI training, the PMBOK®, CMMI and ISO9000. Topics include: Project management (PgM) Planning, project initiation, project execution, monitoring and control, project close-out, risk management, customer relationship management, quality assurance and contract growth. There are about 160 pages of valuable guidance, lessons learned, figures, templates and checklists.

I'm a co-founder and the Chief Knowledge Officer at The Government FreeLance Exchange or GovFlex.com -- the leading online freelance exchange for government contractors & agencies to acquire the services of independent experts. I also founded Celerity Works in 1999 to provide business development (BD) and project management advisory services where I advised over 70 contractors and agencies and coached hundreds of executives, PMs and BD professionals. Prior to that, I was a BD and operations executive for information technical (IT) contractors for 13 years after being an engineer and PM for 15 years.

I've presented numerous knowledge webinars, written hundreds of columns for government business magazines and blogs and implemented the BD and project management process and training program for several government contractors. I also developed the risk management and training process for GSA FEDSIM's major IT acquisitions. I've written several business books including *Winning and Managing*

*Government Business* and, with GovFlex CEO Eric Adolphe, *How to Develop a Winning Small Business Innovation Proposal.*

I hope you find this material useful as you work towards successful project completion!

Sincerely,

Mike

I can be contacted at: lisagor@celerityworks.com.

GOVFLEX™

---

The Government Freelance Exchange (GovFlex.com) matches government agencies and contractors to the best independent experts they need for a project. GovFlex was established to provide leading e-Commerce-based practices to those who serve in government, the military, as well as the businesses and independent experts who support them.

The company was created to streamline the process of finding needed talent quickly at competitive prices, while promoting compliance and ease of use. It is dedicated to enhance competition and reduce overhead the same way that "gig economy" platforms have enabled private sector enterprises and freelancers to work together on a project basis.

GovFlex is a true end-to-end solution that begins with defining the expertise and skills required. The system identifies the most qualified subject matter experts available, helps the client identify the right expert and on-board them for the project. Our service continues through completion and payment for the project, even providing IRS Form 1099 income tax information to the expert at the end of the year. We protect the Security and personal information of the experts and those who seek their services. Our prompt payment processing and handling of related financial information are leading practices in the government marketplace.

# Introduction

This Guide describes the key elements of a generic government project management process. The four main sections correspond to the project lifecycle phases illustrated in this figure:

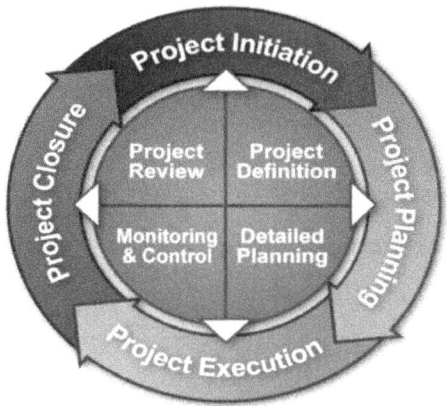

In general, this Guide is also consistent with ISO9001 and contains all the major elements of a quality assurance program for services and solutions provided to government clients. It describes the process used by successful government contractors and some agencies to plan and control the performance of a project from inception to closure. It's also consistent with government contractor best BD practices (see author's *Winning and Managing Government Projects*).

The first two sections deal with **what** a project will produce...in other words, the desired outcome. And with **how** the project will be accomplished. Where practical, the PM should attempt to complete as much of these phases as practical prior to contract award.

I've included into each section and in the appendices over a hundred valuable lessons from my experience managing or reviewing numerous government projects.

## Document Scope

As a project manager (PM), you need two types of training:

- Facts...the "how you do it at your organization"
- Skills...the basic techniques every manager should possess

The four major sections in this Guide focus on the facts...the specific activities, responsibilities, interfaces and resources you need to be successful. They also provide many useful figures, templates and checklists.

The Guide's target audience is newly promoted PMs and more experienced PMs in need of a refresher.

All PMs are encouraged to further develop their PM skills through training from the PMI and DAU Project Management related curriculum.

The Guide covers these PMBOK® *Guide* knowledge areas:

- Project Scope Management
- Project Time Management
- Project Cost Management
- Project Quality Management
- Project Human Resource Management
- Project Communications Management
- Project Risk Management
- Project Procurement Management

There are numerous sizes and flavors of government projects ranging from major services contracts involving multiple labor categories to complex software development to simpler projects comprised of just one or two person tasks. The complexity of a project also depends on whether your organization is the prime or subcontractor. Regardless, it's important to involve other experienced managers in the planning process to get their perspectives to benefit from lessons learned on previous projects.

There is a difference between a Project Manager and a Program Manager. Project Managers are usually responsible for a single project or several smaller projects or tasks usually for a single government agency. A Program Manager is usually responsible for multiple larger projects and might have one or more PMs reporting them. Their span of control might encompass more than one agency. Specific PMI definitions of project and program management are provided in the two following paragraphs.

**Project management** is "...the application of knowledge, skills, tools, and techniques to project activities in order to meet or exceed sponsors' needs and expectations from a project (PMBOK® Guide)." Project management balances competing demands (scope, time, cost, quality, requirements, etc.) throughout the project lifecycle. Limited available resources require the efficient use of contract dollars. Project management helps your organization maintain efficiency by making sure that the right people complete the right tasks at the right time.

A **program** is "...a group of related projects managed in a coordinated way to obtain benefits not available from managing them individually (PMBOK® Guide)." Laws and regulations establish programs for government projects.

This Guide deals primarily with the activities of a **single project** managed by a PM. It focuses on the key activities that the PM is required to perform before and after project award, or when inheriting an existing project. The definitions for any unfamiliar project management terms can be looked up in the Glossary and List of Acronyms.

Your organization's senior management should ensure the PMs authority, responsibility and latitude is put in place so that effective management of the project will result. It's senior management's responsibility to create and foster an environment that facilitates successful project management and execution. This includes being:

- Assigned *responsibility* for the successful execution of the project.
- Given *authority* over the assets, resources, and personnel needed to successfully execute the project.
- *Rewarded* for successful performance.
- Held *accountable* for his or her actions.

This Guide assumes that you will be supported by senior management. But, regardless of the organizational challenges you might face, I'm confident that your efforts to expand and apply the knowledge in this book and elsewhere will pay huge dividends in your future.

# Section I. Project initiation

Project initiation is comprised of distinct activities that, where possible, should be undertaken before contract award. The outputs of the project initiation activities should eventually be documented in the PgM Plan described in Section II, PgM Planning.

These initiation activities include:

1. Understand the contract, technical and staffing requirements
2. Understand project initiation risks
3. Identify the key stakeholders
4. Develop an initial project action plan

It's important to note that the activities described in this section represent an ideal scenario that might not always be possible. For instance, a PM may not have yet been assigned or there might not be sufficient time prior to contract award to accomplish these tasks. For ease of reading, some PM activities that should start, if possible, during the project initiation phase, such as WBS development, schedule development, and risk identification, have been consolidated into the Planning phase, Section II of this Guide.

The following table summarizes the key information in this section:

| | Tasks | Deliverables (Outputs) | References | Interfaces | Key Points |
|---|---|---|---|---|---|
| 1. | Understand the contract, technical & staffing requirements | -Requirements traceability checklist<br>-Risk assessment activity (see Section II) | -Copy of proposal, RFP, teaming agreements | -Capture manager, contracts, project control, subcontracts, Security, HR, facilities, IT<br>-Contracts, Subcontracts | -Understand key elements of contract risk for your contract type (FFP, CPFF, T&M, etc.)<br>-Pay attention to all "shall" statements<br>-Contracts & subcontracts provisions may not be the same |
| 2. | Understand project initiation risks | -Preliminary risk register | -Copy of proposal, RFP, contract (if post-award) | -Managers and Corporate Compliance | -Develop "what-if" scenarios for high probability/high impact risks<br>-Consider technical, cost & schedule<br>-Mitigate major risks before they become issues! |
| 3. | Identify the key stakeholders | -Project stakeholder list<br>-Contact plan | -Stakeholder matrix | -BD, capture, Division Managers | -Identify key influencers & decision makers |
| 4. | Develop an initial project action plan | -Project action item list | -Your organization's organization PM Guide<br>-Risk register | -Proposal, BD, Contracts, Subcontracts, Project control | -Identify responsible party and due date<br>-Regularly status including at project kick-off meetings |

# 1.  Understand the contract, technical and staffing requirements

Understanding the requirements is one of the best project success strategies a PM can employ. If you were not involved in the proposal process, get a copy of the technical and cost proposal. Following contract award, proposal documents will probably be maintained as part of the contracts file.

Compare the commitments and assumptions made in the proposal with your understanding of the Request for Proposal (RFP) and Statement of Work (SOW) requirements. Make a list of any ambiguities and anything you're not sure you can complete. Obtain clarification of these items from your Contracts Administrator. Remember, that a failure to manage to the contract requirements usually results in significant problems including contractor failure to deliver and requirements scope creep.

One key directive for all PMs is "Manage to your contract." This implies having a thorough knowledge and understanding of the contract type. It's also important to recognize that the contract type that was proposed or that is required by the RFP may not be the same contract type that ends up being negotiated with any subcontractors. There are several factors that impact this decision including the type of subcontractor service or product and subcontractor performance risk.

Keep in mind that too many failed government projects are the result of poorly written technical and operational requirements. My standard, after some costly mistakes, became, "If it can't be verified by testing or inspection, then it isn't specific enough." For instance, "The System shall be user-friendly," (a real requirement from a 'he-who-shall-not-be-named' agency specification), is useless. If you accept this kind of ambiguous requirement, you'll have left yourself open to massive arguments with your government client about project acceptance and financial liability.

It's very important for you to understand the full pricing approach behind the cost proposal that was submitted to the government customer. These detailed pricing files were probably maintained in a corporate pricing database. Your responsible Contracts Administrator can assist you in obtaining the necessary pricing files that detail cost and profit factors used to develop the cost proposal. Identify any gaps between what was proposed and your plan. It's not unusual, especially on larger procurements, for there to be a long time between proposal submittal and contract award. Your organization's salary guidelines and overhead costs may have changed and circumstances such as product releases and resource availability may impact your ability to deliver to the proposed schedule. All of these factors will need to be taken into consideration and, where necessary, coordinated with the Contracts Administrator upon contract award.

There are several critical IT support elements that need to be coordinated with your organization's IT Department. These include determining whether your project will be using Government Furnished Equipment (GFE) at the client site or your site, or will require your own organization's IT equipment. Your project's network access, remote access and system administrator support will also need to be identified. In addition, you should make sure to understand which of your IT equipment will be considered General Support Equipment vs. Project Specific Equipment as this will impact your budget and how these items are charged.

There may be facilities requirements that will need to be coordinated with your organization's Facilities Department to ensure you and your staff have the necessary support on day one. Early interaction with your Security team is critical to the successful implementation of any Security requirements, such as clearances, facilities, IT systems, etc., identified in your contract. An understanding of the Security requirements can influence your technical approach and the successful implementation of them will have a positive impact on your project's success.

A significant first step in the PgM Planning process for procurements with sufficient lead time between proposal submittal and contract award is to hold a meeting with the Proposal Manager and the Capture Manager (unless one or both of these was you!) to confirm your understanding of the key elements of the project. This meeting will provide much of the background information necessary to begin development of the PgM Plan.

Here are the key elements of a proposal hand-off meeting:

- Contract type
- Contract duration
- Technical approach
- Deliverables
- Key milestones
- Key personnel
- Cost proposal
- Project risks

- Contract set-up sheet
- Client stakeholders
- Client issues and hot buttons
- Security requirements

The PM is required to create a Document Management Matrix (DMM). The DMM provides details about the control and management of project documents, records, and written work products, such as the current version of the document, storage location, monitoring method, back up method and frequency, access rights, and retention period. The initial records to be included in the DMM are specified below:

- SOW
- Questions and/or clarifications submitted during the proposal phase, and responses received
- Your organization's technical proposal (both written and oral materials)
- Cost proposal assumptions (Revenue and Profit details) and associated projected level-of-effort estimates
- A copy of the Risk Log prepared during the proposal process

Other contract or project records are collected and added as work on the project progresses. Pay attention to all of the "shall" statements (what your organization shall do) and "will" statements (what the government will do) in the RFP. These requirements are usually documented in a proposal traceability compliance matrix. Look to see if there are any ancillary requirements such as special equipment, Security clearances, technology specialists, certifications or unique training. Also, make sure you understand how and when you'll be reporting to the customer.

Recruiting for key staff often begins during the capture or proposal phase for a project depending on the type, priority, and complexity of an acquisition. This is especially true when there is a need for cleared staff or highly specialized key personnel. A roles and responsibility matrix is a good tool to use for this exercise. This is a table that describes the roles and responsibilities of each key member of the Project Team.

It's important to get a copy of any teaming agreements to understand the commitments that may have been made as well as RFP and proposal subcontracting requirements so you can factor these into your staffing plans.

You should then develop a staffing approach including the desired mix of internal assignments vs. new hires. You should ascertain internal staff availability and skill match.

Fill out any necessary contingent hiring requisitions in your organization's recruiting system. This process should then be routed to the Recruiting Department for approval and to commence recruiting search. Follow-up regularly with your manager to make sure needed resources will be made available or employment offers are being made on a contingent basis.

You'll also need to carefully balance staff skill level vs. RFP labor category requirements to manage contract labor costs. All contracts that have labor category requirements need to follow your organization's labor qualification guidelines.

There is most likely an established recruiting process that you should follow to support pre-award contingent and key personnel recruiting and hiring for your project. Keep in mind that the more accurate and complete the requisition form, the higher the probability that recruiting will be able to provide you with qualified candidates. Good screening questions will also help recruiting narrow the search.

Your online schedule needs to be up to date at all times to facilitate a smooth and timely interview process.

## 2. __Understand project initiation risks__

Risk management is discussed in detail in Section II. However, here are some negative risk scenarios that you should keep in mind as you prepare for contract award:

- Not involving all key project stakeholders
- Not understanding who the real client is
- Vague objectives or requirements or poorly defined deliverables
- Vague or nonexistent role and responsibility definitions
- Incomplete, inaccurate or unreasonable schedules
- Changes to the requirements baseline
- Lack of configuration and quality assurance management process

Some risk scenarios consist of positive scenarios like:

- Receive additional tasking
- Delivery date is moved out by the client

An inability to make sound decisions is often another source of major project risk. It has been my experience that too many PMs (and executives) make significant decisions in the hallway without a thorough analysis and discussion with all interested parties. Avoid this! Instead, adopt a sound decision-making approach that identifies success factors and rates alternatives against these factors. Then communicate the decision.

While delaying a decision can be demotivating, costly and unproductive, at other times it can allow you to gather more information and consider new alternatives.

## 3. Identify key stakeholders

It's hard to overstate the importance of relationships to the successful management of a project. You need to identify the key individuals who are project decision makers or influencers. These people are not always readily apparent. For instance, many IT projects are actually funded by a client's operational unit whose buy-in is needed both during the requirements phase and to obtain delivery acceptance. Stakeholder analysis is covered in more detail in Section II, PgM Planning and Section III, PgM Execution.

## 4. Develop a project action plan

Every PgM Plan should contain an action list that describes the action, the responsible party, an expected completion date and a status. This action list is probably the single biggest risk mitigation step you can take. You should coordinate this list with all interfacing groups and maintain the list throughout the project lifecycle.

# Section II: Project planning

*"I always wanted to be somebody, but now I realize I should have been more specific."* – Lily Tomlin

Successful management of your organization's projects starts with a solid project definition and kick-off activity. This involves the following tasks:

1. Requirements Baseline
2. Work Breakdown Structure (WBS)
3. Financial (cost estimate) baseline
4. Staffing requirements
5. Schedule baseline
6. Contracts baseline
7. Subcontracts baseline
8. Facilities
9. IT support
10. Security
11. Material procurement
12. Authority (stakeholder) matrix
13. Configuration change control
14. Quality assurance and control
15. Risk management
16. Action planning
17. Kick-off meetings
18. Project management (PM) plan

The outputs of the first 17 planning tasks should be documented in a PgM Plan (task 18) and maintained in a Project Notebook or, where available, by referring to an online project folder. It provides a vehicle for defining the major elements of the project and also serves as a valuable transition document if a new PM comes on-board. It's critical that you involve all the internal organization and external customer stakeholders in this process.

As the PgM Plan is a living document, you should regularly revisit and update it to reflect current project status and to make sure you haven't neglected an important aspect of the plan.

The following table summarizes the key information contained in this section:

| | Tasks | Deliverables (Outputs) | References/ Tools | Interfaces | Key Points |
|---|---|---|---|---|---|
| 1. | Requirements Baseline | -Requirements traceability checklist -Scope description | -Your organization's Authority Matrix | -Contracts, project control, subcontracts, Security, HR, facilities, IT | -Pay attention to all "shall" statements -Look for gaps between proposal and RFP -Review staffing, deliverable, IT, Security, reporting Requirements |
| 2. | Work Breakdown Structure (WBS) | -WBS structure in diagram -WBS dictionary | -MIL STD 881 | -Project Team | -Use to build consensus -Consider best approach: deliverable, process or activity based -Status/completion is measurable -Duration & cost can be estimated Work elements are definable- -Where necessary, break tasks down to ~80 hours (but not too small) |
| 3. | Financial (cost estimate) baseline | -Usually an Excel project budget spreadsheet -Project cost summary -ABC Financial System contract & financial reporting structure | -Your organization's Financial System -PM Dashboard -FAR Part 16 | -Pricing, program controller, finance analyst/Biller, contracts, subcontracts | -Estimates should be built bottom-up (engineering or business based) -Request copy of cost proposal from Contracts (by PL#) -Check for specific contract financial reporting requirements -Consider management reserve for high risk projects |
| 4. | Staffing requirements | -Your organization's organization chart -Roles & responsibilities matrix -Staffing profile in PgM Plan -Resource loading spreadsheet | -Your organization's Recruiting Procedure -Interview summary form -Wage & Salary Class guide | -Project planning, recruiting, subcontracts, Security | -Make contingent offers (pre-award) -Pay attention to labor rates & skill requirements -Get a copy of teaming agreements to under commitments -Keep online interview schedule up to date -Never cancel an interview (find a back-up) -Fill out interview summary form |
| 5. | Schedule baseline | -Milestone charts, activity lists, Gantt charts, activity networks, critical path networks, PERT charts | -Microsoft Project or equivalent | -Project Team | -Place where WBS, cost estimations & staff assignments come together -Options to correct estimated schedule exceeding client requirement: re-negotiate, add additional resources and/or reduce scope of project -Accept reasonable challenge |

| Tasks | Deliverables (Outputs) | References/ Tools | Interfaces | Key Points |
|---|---|---|---|---|
| 6. Contracts baseline | -Signed contract | -Your organization's Financial System <br> -Your organization's Contracts Start-up Procedures | -Contracts, Subcontracts, Project Control, HR, Security, IT, Facilities | -Obtain clarification of any ambiguities <br> -Understand your contract type (FAR Part 16) <br> -GFE or your organization's servers? <br> -IT equipment General Support Equipment or Project Specific Equipment? <br> -Contract & subcontract clauses may not be the same |
| 7. Subcontracts baseline | -SOW & purchase order (ABC Financial System) <br> -Signed subcontract | -Your Organization's Financial System <br> -Contract <br> -Teaming agreement | -Subcontracts <br> -Contracts <br> -Procurement | -Must have signed subcontract prior to starting work <br> -All subcontract commitments and scope changes must be approved by a VP and the Contracts Director <br> -Required to have full competition to select vendors |
| 8. Facilities | -Facilities requirements defined in PgM Plan | | -Facilities | -Try to communicate facility needs as soon as possible |
| 9. IT support | -IT requirements defined in PgM Plan | -GFE identified in contract & the proposal | -IT Department | -Carefully assess all IT needs to support Project Team & delivery |
| 10. Security | -Security requirements documented in the contract | | -Group Security Team | -Initiate interaction with Security at beginning of program <br> -Security can assess your requirements, determine ability to support, provide current capabilities and timelines for support <br> -Manage and provide day-to-day Security support to your program |
| 11. Material procurement | -Bill of Materials <br> -Purchase order | -Purchase request (PR) | -Procurement <br> -Proposal manager <br> -Contracts | -Work closely with Procurement Specialist <br> -Identify everything on bill of materials <br> -PR to approval can take up to three weeks |
| 12. Authority (stakeholder) matrix | -Authority matrix <br> -Influence map | | -Project Team | |

| Tasks | Deliverables (Outputs) | References/ Tools | Interfaces | Key Points |
|---|---|---|---|---|
| 13. Configuration change control | -Change process defined in PgM Plan<br>-CMM tool for change control<br>-Cost accounting processes including EVM | -Change control tool | -Quality Director<br>- | -Include: traceability, prototyping, modeling, impact analysis, reviews |
| 14. Quality assurance and control | -Quality process defined in PgM Plan | | -Quality Director | -Specify use of quality control processes including conformance to work processes, verification & validation, joint reviews, audits & process assessment |
| 15. Risk management | -Risk process defined in PgM Plan<br>-Risk register (log) | -ISO 16085<br>-PMBOK | Group Program Execution Authority | -Assess business, technical, schedule, resource and programmatic risks for impact & probability<br>-Maintain a risk register (log) |
| 16. Action planning | -Project action item list | -Risk log | -All key stakeholders | -Maintain and regularly status action item list |
| 17. Kick-off meetings | -Meeting agendas<br>-Meeting actions<br>-Internal team kick-off, Project Team kick-off w/subcontractors & client kick-off<br>-Project success criteria | -Meeting agenda | -All key stakeholders | -Invite all stakeholders to ensure effective integration & communication<br>-Pass out agenda in advance<br>-Begin & end on time<br>-Stick to the agenda<br>-Draw people out (silence isn't consent)<br>-Record decisions & action assignments |
| 18. Project management plan | -PgM Plan | | -All stakeholders | -PgM Plan required on every project, tailored as appropriate<br>-Use Appendix A - template for detailed example<br>-Should be a living document<br>-Transition tool for new PM |

1. **Understand the requirements**

As discussed in Section I-1, it's critical to the success of the project that, as the PM, you fully understand the scope and requirements of your project. Not only is it smart management and business, but it also helps mitigate project risks.

Thoroughly review the entire contract including any negotiated items to ensure that the approaches to be taken are consistent. Where discrepancies are found, document them, track to closure, and communicate all intended changes to management for review and approval.

On development projects, evaluate any technical requirements to make sure they are specific enough to allow you to pass a government acceptance test. If necessary, provide further requirements definition to be finalized with the client during the Project Planning phase.

If appropriate, meet with the cognizant business development or sales executive to review: the status and outcomes of past discussions with the customer, your understanding of the customer's issues, needs, and environment; and any assumptions made in developing the project proposal.

Document your project's scope and requirements baseline. 'Shall' statements identify the technical, schedule and cost reporting requirements. Record them, as well as:

- Special equipment
- Unique personnel skills, certifications, and training
- Security requirements (clearances, facilities, and equipment)
- IT support/equipment
- Discrepancies

2. **Work Breakdown Structure (WBS)**

A preliminary project work breakdown structure (WBS) will provide you with a framework that defines the way the work elements will be organized to ensure that all the necessary activities and deliverables have been considered. By developing an initial version of this before contract award, you will be prepared to start on day one. The start of any project is of great importance. It's the reason I will be spending a little more time on this topic.

Development of the final WBS should be a team activity to build consensus and understanding of the scope and work plan necessary to complete the project.

There are several approaches to define a WBS...for instance, by deliverable, process (function) or by activity. Regardless, it should clearly identify the necessary tasks and should be used to develop schedules, cost estimates, skill sets, and project risks.

*Deliverable-based WBS example:*

| |
|---|
| **1.Build a computer based training course**<br>**1.1 Module one**<br>1.1.1 Plan<br>1.1.2 Implement<br>1.1.3 Review<br>1.1.4 Accept<br>**1.2 Module two**<br>1.1.1 Plan<br>1.1.2 Implement<br>1.1.3 Review<br>1.1.4 Accept<br>**1.3 Module three**<br>1.1.1 Plan<br>1.1.2 Implement<br>1.1.3 Review<br>1.1.4 Accept |

*Pros:*

- Maps directly to deliverable items
- Provides visibility into costs at the product level
- Easy to explain to customers

*Cons:*

- Cost accounts get spread throughout the WBS

*Recommended use:*

- Delivery of infrastructure (i.e., HW and SW) components or when low labor costs are lower than material costs

*Process (function)-based WBS example:*

---

**1.Build a computer based training course**
**1.1  Plan**
1.1.1  Develop plan
1.1.2  Review with the boss
1.1.3  Finalize plan
**1.2  Implement**
1.2.1  Identify requirements
1.2.2  Design the system
1.2.3  Develop the contents
1.2.4  Finalize the training material
**1.3 Accept**
1.3.1  Review with users
1.3.2  Incorporate changes
1.3.3  Release new course

---

*Pros:*

- Maps directly to defined your organization's's practices/processes

- Reusable across multiple projects (i.e., needs only be defined once then tailored for each specific instance)

- Facilitates schedule planning & tracking

- Supports rapid response to customer needs

*Cons:*

- Actual product component costs not readily visible

- Requires additional customer interaction to convey benefits of this approach

*Recommended use:*

- Delivery of repeatable services (e.g., consulting services, integration services, etc.), configured packaged solutions (e.g., ERP system, CRM system, etc.), or whenever schedule is of paramount concern

So, clearly a WBS can be created in different ways to emphasize different views of the project. Some customers supply their own product-focused WBS. As the PM, you may need to gather other Department views of the project so that they can understand the budgets relative to your project so they are consistent with the overall cost components. Other WBS structures might be created to view materials and labor separately or to separate various other categories of expense.

The need for different views of the work can and often does lead to the implementation of multiple WBSs. In this situation, a mapping across WBSs must be maintained. While the WBS is used to breakdown the work and budget into more easily understood pieces, it also serves as a device to communicate within the project the work to be done and the budget assigned to that work. Too many WBSs on one project can create extra work and cause confusion. For small projects, a single WBS view aligned to meet the Project Team's needs is best.

It's sometimes easier to picture the relationship between the various work elements in a graphical representation such as the one pictured here:

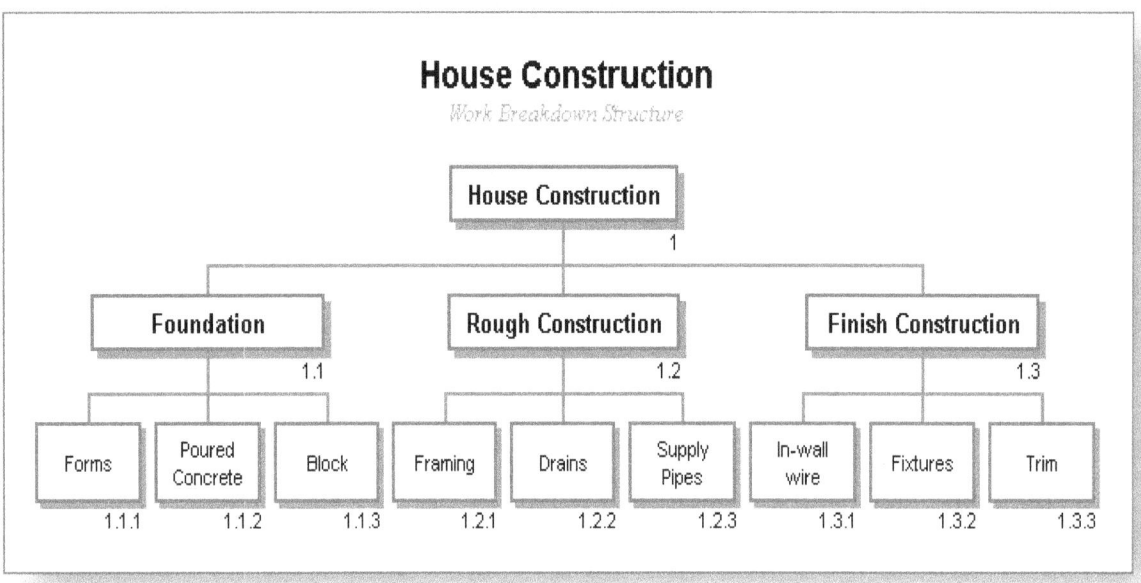

How the contract cost elements are defined will depend on whether the project is a development contract or a labor-based contract, how you choose to manage the various project elements, and the project progress reporting requirements.

Each work element or activity should have a measurable milestone such as a milestone completion review and, if possible, a definable output or deliverable. The start/end events should be clearly defined. The activity duration & cost should be easily estimated. Work assignments should be independent and definable.

A deliverable (product)-based WBS is usually preferred if the client wants discrete deliverables/products. However, this may not be possible on certain types of labor support contracts.

This table represents part of the design phase of a system development project:

| WBS | CLIN 0002 Design Phase Elements |
|---|---|
| 2.1.1 | System data modeling |
| 2.1.1.1 | Logical data model |
| 2.1.1.1.1 | Validate logical data model document, deliver to customer |
| 2.1.1.1.2 | Customer review logical data model |
| 2.1.1.1.3 | Finalize and deliver logical data model |
| 2.1.1.2 | Physical data model |
| 2.1.1.2.1 | Validate physical data model document, deliver to customer |
| 2.1.1.2.2 | Customer review physical data model |
| 2.1.1.2.3 | Finalize and deliver physical data model |
| 2.1.1.2.4 | Perform initial baseline of data model |
| 2.1.1.2.5 | Perform initial analysis of lookup values of data model |
| 2.1.1.2.6 | Populate initial lookup values in data model |

This illustrates the decomposition of two major tasks into subtasks. These subtasks are discrete activities that can be measured and, if required, for which earned value can also be tracked.

A preliminary project WBS provides you with a structure that defines the tasks and/or work that needs to be performed to meet project objectives. It organizes and defines the total scope of the project.

A WBS should decompose the project activities to successively lower levels of detail but no further than necessary to assist with the management of the project. It's important to understand that this process is about defining how you'll do the job not just a regurgitation of the SOW.

Some projects might require earned value management. In these cases, you'll need to factor in earned value discrete work packages as WBS elements. Use an earned value tracking tool to set-up and track earned value management status.

The WBS will also need to reflect whether you're planning a fixed price product, a level of effort or a defined task project. Finally, your WBS must also take into consideration quality review activities.

Don't overwork the WBS. It should be only as detailed as necessary to describe what you need to manage the project. Keep in mind the burden being placed on your organization's support staff of dealing with too many work and cost elements. If a WBS activity doesn't possess all five of the attributes illustrated below, then decompose or consolidate it further until the answer to all the questions is YES:

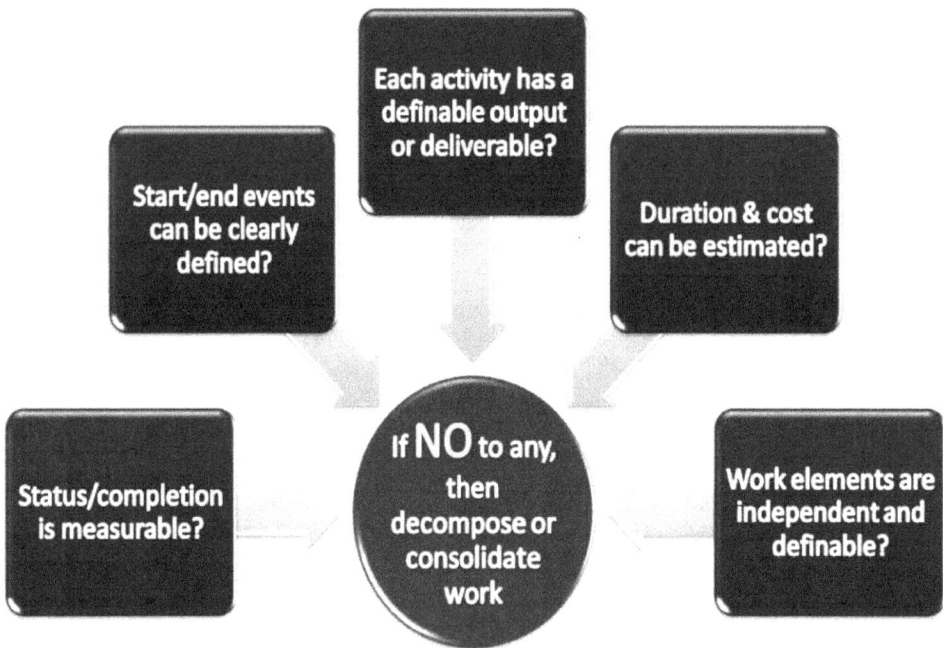

Each work element or activity should have a measurable milestone such as review complete and, if possible, a definable output or deliverable.

The start/end events should be clearly defined. The activity duration & cost should be easily estimated. Work assignments should be independent and definable.

One final point – be certain to check all the special requirements in the RFP. For instance, many Department of Defense contracts specify *Military Standard 881* that gives specific direction regarding WBS development.

## 3. <u>Financial (cost estimate) baseline</u>

Make sure you understand your cost proposal -- including the year one, two and three pricing strategy since you'll have to manage to this proposed budget. You can request proposal copies from the Contracts Department. Where possible, you should support the proposal activities to validate pricing estimates, recognizing that in many cases you won't have begun until after contract award.

*Financial rate terminology*

Some of the common terms used to refer to labor rates and cost structure are:

- A labor rate (sometimes called a direct rate) is a term that needs to be modified with either the word unburdened or burdened to make complete sense. An unburdened labor rate is the hourly rate being paid to an employee.
- A sell rate is the hourly labor rate fully burdened with G&A, overhead and fee that your organization is proposing to the government for a specific labor category. See your Project Control Analyst to get the financial related overhead structure you should be incorporating into your budget estimate.
- A billable rate or fully burdened rate is the final negotiated hourly labor rate on the approved contract.
- A wrap rate is a factor that when multiplied by an hourly labor rate equals the final fully burdened rate minus profit.
- An operating unit is a collection of similar work within a defined organizational unit that applies the same wrap rate to hourly labor rates. In other words, an operating unit, sometimes called a cost center, has the same overhead & G&A structure.

The table on the next page is an example from the Transition-in Phase WBS of a project in an Excel spreadsheet form. This is only a part of the overall WBS but illustrates how you can document the labor hours for each of the WBS subtasks and then calculate corresponding budget dollars by multiplying the hours times the fully burdened labor rate. A general rule of thumb is that a task should be broken down until it's below 80 hours in order to adequately monitor and control that task.

| WBS | Hours | CLIN 0001  Transition-In Phase Events | Budget |
|---|---|---|---|
| 1.1.1 | 48 | Vendor coordination (assume 12 meetings) 4 hours each | $4,202.88 |
| 1.1.2 | 32 | Project management & Project artifact transition | $2,801.92 |
| 1.1.3 | 4 | IBR Milestone | $350.24 |
| 1.1.4 | 30 | Development environment setup | $2,626.80 |
| 1.1.5.1 | 80 | Review Requirements Documents | $7,004.80 |
| 1.1.5.2 | 66.5 | Revise Task Estimates | $5,822.74 |
| 1.1.5.3 | 24 | Requirements Elaboration meetings (Assume 6 meetings) 4 hours each | $2,101.44 |
| 1.1.5.4 | 32 | User Interface design meetings (Assume 8 meetings) 4 hours each | $2,801.92 |
| 1.1.6 | 71 | Updated Supplemental Specifications Document | $6,216.76 |
| 1.1.7 | 80 | Updated Glossary, Acronym List, and Business Rules | $7,004.80 |
| 1.1.8 | 70.5 | Updated Requirements Traceability Matrix | $6,172.98 |
| Total | 538 | | $47,107.28 |

The following table is an example of a cost summary organized by contract line items in a sample project WBS in an Excel spreadsheet form:

| CLIN | Hours | Title | Cost Estimate |
|---|---|---|---|
| CLIN 0001 | 1347 | Transition-In Phase | $134,485 |
| CLIN 0002 | 5042.7 | Design Phase | $413,954 |
| CLIN 0003 | 36890.7 | Development and Test Phase | $3,020,318 |
| CLIN 0004 | 2249 | Implementation Phase | $199,280 |
| CLIN 0005 | 4614 | Operations & Maintenance Phase | $437,098 |
| CLIN 0006 | 6970.1 | Program Management Activities (throughout all phases) | $795,495 |
| CLIN 0007 | - | Special Projects (Optional) | $400,000 |
| | Total | | $5,400,630 |

Check the contract for specific financial reporting requirements. The Department of Defense specifies the *CSSR system* with associated Contract Data Requirement Lists (CDRL) and Data Item Descriptions (DID). Civilian agency contracts have a variety and, unfortunately, inconsistent set of financial requirements.

*Estimation*

Note that you should understand the two basic proposal cost estimating methods...top-down which is based on similar projects and bottom-up which is developed by estimating each cost or WBS element of a project.

Specify the estimated cost for direct project personnel, and include as appropriate the costs for travel, meetings, computing resources, software tools, special testing, simulation facilities and administrative support. By comparing a top-down estimate with your own bottom-up estimate, you can identify any areas of staff or schedule risk and, in some cases, the need for additional resources.

Next, with the assistance of Project Control or Finance Department staff, create an Excel spreadsheet (or use an approved organizational project control spreadsheet/tool) to track weekly or monthly labor hours by staff member fully burdened billing rate. You should make sure the final contract burden or overhead and profit have been included in the hourly labor rates in your spreadsheet. The selection, development, and approval of project-level planning and management tools probably occurs at your organization's corporate level.

Ultimately, you're the person responsible for managing to your budget. On labor based contracts, it's important to balance staff assignments with negotiated labor categories. Clearly this is also critical on fixed price contracts. In some cases, the cost proposal will have been constructed to achieve a lowest credible cost and will not include a management reserve budget amount to accommodate unplanned schedule delays or technical deficiencies. This management reserve is a certain percentage of the overall budget that is set aside for unplanned contingencies. In other instances you might be able to build a management reserve dollars into your budget estimate as a risk mitigation method.

Pay particular attention to labor rate changes if there has been a significant lapse of time since proposal submittal. Also, study teaming arrangements with any subcontractors so that you can factor their labor and other costs into your budget. In some cases, subcontractor work share may decrease over time. Or, in other cases, the RFP may require a specific percentage of work be allocated or set aside for different categories of small business such as 8(a), veteran, minority, women and native American owned.

On fixed price delivery and performance-based contracts, your cost tracking will be for internal cost management to make sure you manage to your budget. The actual invoices will reflect pre-determined fixed price deliverables.

On labor hour contracts, your cost tracking will be reflected on the invoices you'll be submitting through your Biller to the client for approval and payment.

*Assigning Charge Numbers in a Financial System*

Upon award, your Project Biller (or designated Finance staff member) will set up the contract in the Financial System according to the internal cost accumulation requirements and billing requirements in the contract. You should discuss your WBS with your designated Contract set-up personnel, prior to award if possible, so the coding structure in the Financial System is consistent with your WBS. This will make it much easier for you to use the resulting financial reports to manage your project.

Every contract will be assigned a contract #, and, usually, at least one contract line # with funding and period of performance, and at least one project and activity ID. A contract often has multiple tasks. Each of these might have a unique project ID. Here is an example of a project structure that has been set-up to be consistent with the project WBS.

- **Contract #H003** (Department of Transportation – Task Order #5)
- **Project ID H003-300** (specific project or task under this contract)
    - **Contract Line 1 - Labor - $100K**
    - **Suffix** (for the labor contract line cost accumulation)
      30 - Direct Labor - Customer Site
      40 - Direct Labor – Contractor Site
    - **Contract Line 2: Travel - $100K**
      50 - Employee Travel
    - **Contract Line 3: ODC - $100K**
    - 55 - Other Direct Costs

There are usually four key individuals* who have a specific role in the development of the project financial baseline.

- PM: provides project budget and financial setup information
- Project Control Analyst: Assists PM with setting up initial project budget and spreadsheet
- Finance Analyst/Accounts Receivable Specialist: sets up new cost, time, labor and billing structures in ABC Financial System
- Subcontracts Administrator: sets up new vendors and subcontractors in your organization's Financial System

Your ability to work smoothly with each of these individuals will greatly contribute to your project's success.

*Note: Due to unique organizational process differences, these functional roles may be performed by individuals with job titles not listed here.*

## 4. Staffing requirements

Assembling your Project Team is one of your most important tasks…getting the right people with the right skills at the right time. This is especially true when there is a need for cleared staff or highly specialized key personnel. So, it's important to devote as much time to this as you do to developing your work plan and budget.

This activity usually begins with the development of a project roles and responsibilities matrix; basically a table describing the role of each key member of the Project Team. This can most effectively be accomplished using an Excel spreadsheet.

Provide a detailed itemization of the resources to be allocated to each major work activity in the project WBS. Specify the numbers and required skill levels of personnel for each work activity. You must also comply with any labor qualification requirements in your contract.

It's important to get a copy of any teaming agreements to understand the commitments that may have been made as well as RFP and proposal subcontracting requirements so you can factor these into your staffing plans.

You should then develop a staffing approach including the desired mix of internal assignments vs. new hires. Then, ascertain internal staff availability and skill match. Be sure to balance personnel qualifications and skill levels with contract labor requirements and your budget.

Once you have identified your staffing needs, follow the Recruiting Department's defined process to ensure your recruiting needs are met. In parallel, you should also notify your human resources (HR) Department for possible internal fulfillment. Additionally, HR/Recruiters will work with your Security team to ensure pre-screening is completed and clearance processing initiated as part of the hiring and on-boarding process. (See section II-10 for more information.)

Fill out any necessary hiring requisitions in the recruiting system. This figure illustrates a typical recruiting process:

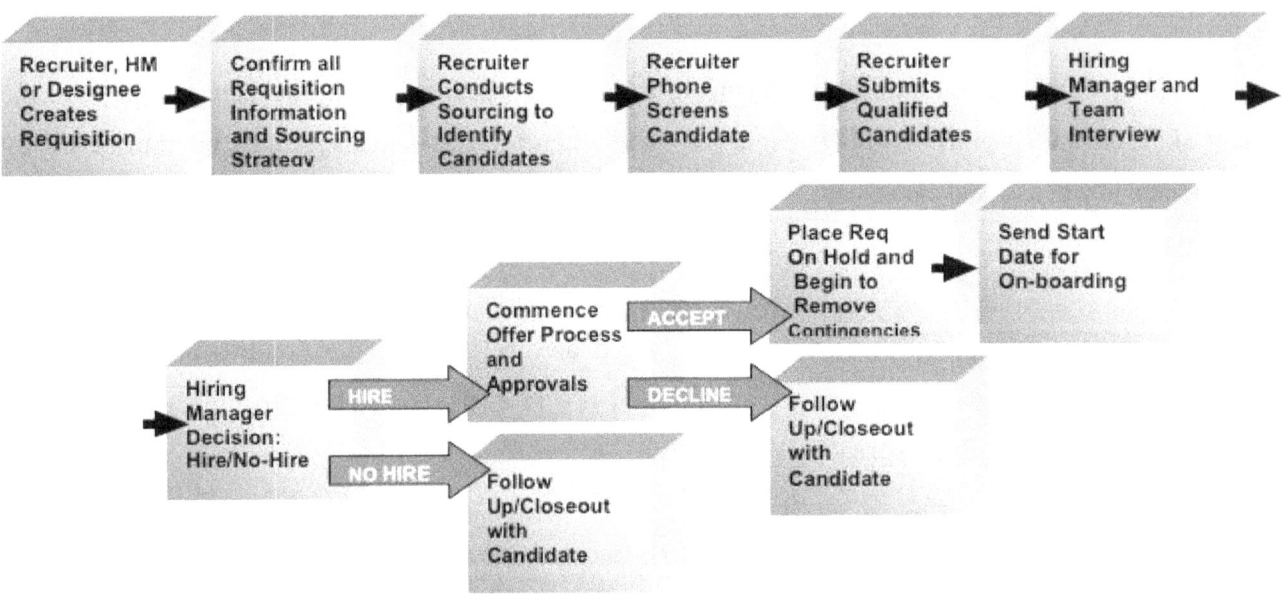

You'll also need to carefully balance staff skill level vs. RFP labor category requirements to control contract labor costs. It's extremely important that every job offer conform to your organization's Wage and Salary Guide as contractors are regularly subjected to government labor qualification audits.

Remember to use similar labor category job titles between prime and subcontract requisitions. Coordinate these with your recruiter.

Keep in mind that the more accurate and complete the requisition form, the higher the probability that recruiting will be able to provide you with qualified candidates. Good screening questions will also help recruiting narrow the search.

Your online schedule needs to be up to date at all times. Try to support a smooth and timely interview process. Conduct your interviews as soon as possible and coordinate any contingent offers with your recruiter.

## 5. **Schedule baseline**

A project schedule may or may not have been required during the proposal process. However, if possible, a high-level schedule should be developed as part of the PgM Plan prior to award. It should be consistent with RFP requirements and the major elements of the initial WBS. As with the other steps in this pre-award Section, it might not be possible to develop the schedule until after contract award.

The WBS, cost estimations and personnel assignments all come together to create a project schedule.

An effective schedule should show the interdependence of project tasks. It should identify project duration and reveal the project's critical path. This will expose project risks that can then be mitigated. Also, don't forget to factor Facility and IT needs into the schedule.

Create a list of tasks that need to be carried out for each deliverable. For each task, identify the type of resources and amount of effort (hours or days) required to complete the task.

Once you have established the amount of effort for each task, you can work out the effort required for each deliverable and an accurate delivery date. Update your deliverables section with the more accurate delivery dates.

The next step is to network the project activities together to define the predecessor (input) relationships and successor (output) relationships between activities to identify the project critical path. A simple example is shown on the right. This will help you understand where the highest degree of schedule risk exists on your project.

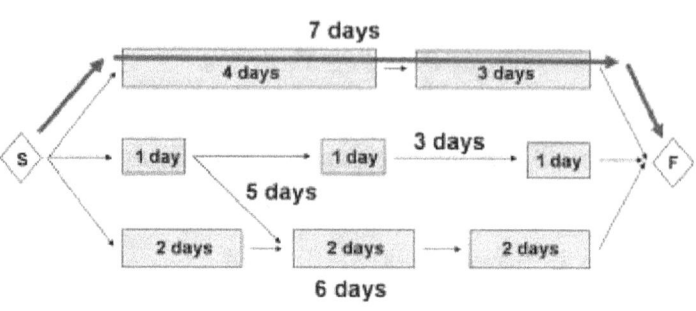

At this point in the planning you should use a scheduling tool to create your project schedule. The following Microsoft Project example won't be of much use but does illustrate the complexity of some development project schedules!

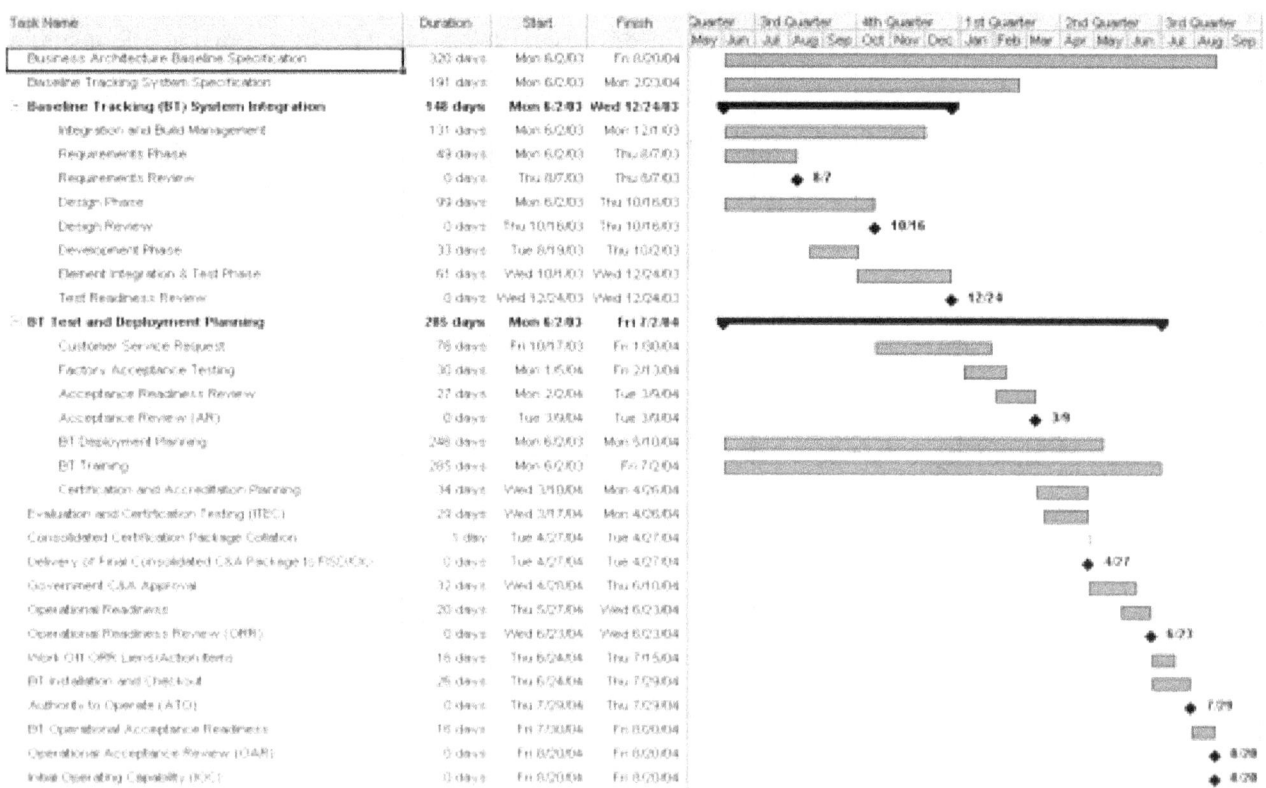

Input all of the deliverables, tasks, durations and the resources for each task. You can explore various schedule planning and reporting options including:

- Milestone charts
- Activity lists

- Activity Gantt (bar) charts
- Activity networks, critical path networks (project critical path)
- PERT charts

A common problem discovered at this point is when a project has an imposed delivery deadline from the sponsor that isn't realistic based on your estimates. If you discover that this is the case you must contact the sponsor immediately. The options you have in this situation are:

- Re-negotiate the deadline (project delay).
- Employ additional resources (increased cost).
- Reduce the scope of the project (less delivered).

Use the project schedule to justify pursuing one of these options.

Depending on the nature of the project, you might also need to adjust the schedule to meet the contractual project end date. However, keep in mind that you need to accept a reasonable management challenge of getting more done for less. But, do try to avoid adopting a schedule that is unsupportable by the available resources.

## 6. **Contracts baseline**

Both you and the Contracts Administrator should read the contract carefully and make note of any discrepancies or areas of concern.

- Verify contract documents against RFP and proposal.
- Check that clauses, terms and conditions, assumptions are the same as RFP and your proposal.
- Check payment terms (EFT or check).
- Verify amounts (labor, travel, other direct costs).
- Verify contract type (FFP, T&M, etc.).
- Confirm key personnel requirements.
- Check for any contract options.

Upon verification that all parts of the documents are as requested/bid, the Contracts Administrator will negotiate with the government Contracting Officer and then sign the final contract.

You should then receive a copy of the fully executed contracting documents back from your Contracts Administrator. It's important to note that only the Contracts Administrator is authorized to commit company resources beyond the scope of the contract.

There are several different contract types, each with its own advantages and disadvantages. Your project management strategy needs to take these into consideration. Familiarize yourself with the *FAR Part 16* (Acquisition.gov). It describes the federal regulations for each the contract types.

You need to be conscious of the various clauses in your contract. In the case of task or delivery orders, there are usually additional clauses in the basic contract. Most of the clauses are described in the Federal Acquisition Regulation.

In general, there are two major types of Federal contract clauses: Standard Federal Contract Clauses and GSA Contract Clauses. For contract vehicles like GWACs and IDIQs, you must adhere to clauses explicitly written in the task order contract and the overarching contract. State and Local projects usually have their own unique standard and project-specific clauses.

**Firm Fixed Price (FFP).** The contractor is paid a price for performing the work and must complete the work to receive this pre-established price regardless of the actual cost incurred by the contractor:

1. If the costs are greater than the price the contractor will suffer a loss.
2. The lower the costs the greater the profits.
3. If the work isn't satisfactorily completed on time, the contractor will be liable for breach of contract. This could entail the contract being terminated for default (failure to perform) which may result in the contractor being held liable for any additional costs incurred by the customer to re-procure the contract items from another contractor.

**Fixed Price Incentive (FPI).** A fixed-price incentive contract provides for adjusting profit and establishing the final contract price by application of a formula based on the relationship of total final negotiated cost to total actual cost. It's a compromise between a firm fixed-price arrangement and a cost reimbursement one. The major FPI elements are defined as follows:

| *Target Cost* | The amount against which final costs are measured. |
|---|---|
| *Target Profit* | The profit for target cost at target performance. |
| *Target Price* | The sum of target cost and target profit. |
| *Ceiling Price* | The maximum dollar amount for which the customer will be liable. |
| *Sharing Formula* | An arrangement for establishing final price, expressed as a customer/contractor share ratio, e.g. 60/40. |

**Cost Plus Award Fee (CPAF).** This is a cost-reimbursement contract in which the fee is earned in part upon satisfactory performance. The fee typically consists of (a) a base amount (which may be zero) fixed at inception of the contract and (b) an award amount that the contractor may earn in whole or in part during performance. The amount of award fee to be paid is subjectively determined by the customer in regard to contractor performance relative to the criteria stated in the contract. This determination is made unilaterally and isn't subject to dispute.

**Cost Plus Fixed Fee (CPFF).** The negotiated fee is fixed at contract inception and doesn't vary with actual cost. (It may be adjusted, however, as a result of changes in the work to be performed.) The contract may take one of two basic forms - completion or term.

- The completion form describes the scope of work by stating a definite goal and specifying an end product. The completion form is preferred whenever the work can be defined well enough to permit the development of reasonable estimates to complete. The completion form requires the contractor to complete and deliver the specified end product (e.g., a final report) within the estimated cost, if possible, as a condition for payment of the fixed fee.
- The term form describes the scope of work in general and obligates the contractor to devote a specified level of effort for a stated time period. The term form shall not be used unless the contractor is obligated to provide a specific level of effort within a definite time period. Under the term form, if performance is deemed satisfactory, the fixed fee is payable at the expiration of the agreed-upon period. (The contractor must provide a certificate that the specified level of effort has been expended.)

**Cost Plus Incentive Fee (CPIF).** A cost-plus-incentive-fee contract is a cost reimbursement contract that provides for an initially negotiated fee (target fee) to be adjusted later by a formula (share ratio) based on the relationship of total allowable cost to target cost. This contract type includes the following elements:

- Target cost
- Target fee
- Share Ratio
- Minimum fee
- Maximum fee

After contract performance, the fee payable to the contractor is determined in accordance with the share ratio.

**Task Order Contracts.** Under a task order contract vehicle, a firm quantity of services or supplies isn't pre-specified. Instead, this type of contract provides for the issuance of delivery or task orders for the delivery of supplies during the contract. The orders become the firm contract obligation. These contracts are commonly referred to as IDIQs. (Indefinite Delivery, Indefinite Quantity) contracts and BOA (Basic Ordering Agreements).

The IDIQ contracts contain a "minimum" order provision that establishes the "consideration" necessary for contract information. BOAs do not commit the customer to procure a specific quantity of services or supplies and therefore are "agreements" in lieu of "contracts" but often establish such things as labor rates. They are often used to purchase management and professional services, studies, analysis and evaluations, and engineering and technical services as well as supplies.

The contractor proposes the particulars for each task order. These proposals may be competitive or sole source depending on the particular contract/agreement. The customer attempts to award multiple task order contracts except when there is only one capable contractor or the cost of administration is prohibitive. The customer believes that competitive pressures result in better price/terms.

**Time and Materials (T&M) Contracts.** The T&M contract provides for payment for direct labor hours at a specified fixed hourly "wrap" rate that includes profit. Materials (i.e., other direct costs and subcontracts) are cost reimbursable and are paid at cost (without profit). A contract-ceiling price is established at award.

**Commercial Contracts.** Commercial contracts may be Firm-Fixed-Price or Time & Materials type contracts. In certain situations, commercial customers may establish a Basic Ordering Agreement with fixed labor rates. The actual labor categories and number of hours are separately negotiated for each order under a BOA. FFP contracts include firm specifications on hardware deliverables. Sometimes schedule penalties are included in a contract to provide an incentive for your organization to maintain project schedule.

Warranties - Usually a one or two-year replacement warranty is the minimum. Warranty reserves may need to be set up to cover the cost of replacements and provide product support.

Risk - Commercial contracts are inherently riskier while also offering potentially higher profit margins. Risk isn't bad, but it must be understood and mitigated. Backup or Contingency plans are essential in successfully dealing with risk. Also, the higher the risk the more the need to make sure management understands the contract.

Intellectual Property - If a customer pays for the development of a product, it's not unreasonable to provide the customer with some sort of exclusive sales rights for their specific components. In general, your organization retains all ownership and other rights to the design.

Indemnification - If a component is designed by your organization and the Intellectual Property rights reside with your organization, customers will want your organization to provide them with some sort of patent infringement protection from third party lawsuits. This isn't unreasonable, but the cost risk should be capped at a value acceptable to business unit management. This typically means that the cost risk is limited to the value of the contract.

International Considerations. Proper and timely review of export licensing requirements is critical to your business. Export of technology, know-how, and hardware are a major concern in international contracts and even in domestic contracts with foreign national employment, VISAs, visitors, consultants, or access to programs. Your organization might be involved in many areas and technologies that are controlled because of DOD and U.S. Government interests. Export laws were enacted to control the flow of vital U.S. technology for national Security purposes, implement foreign policy, control U.S. technology, and to halt the proliferation of weapons of mass destruction.

The Department of State (DOS) is charged with controlling exports and defense services under the International Traffic in Arms Regulations (ITAR), which contains the U.S. Munitions List (USML). The ITAR restricts the export of USML articles and services that may be used for military applications. The Department of Commerce (DOC) implements and enforces the Export Administration Regulations (EAR), which contains the Commerce Control List. The DOC administers the export of commercial and dual use items, and their associated software and technical data.

In some cases, exports governed by the EAR require prior written assurances from the foreign customer that the export not be further released to other countries. Specific statements may also be required on the shipping documents. The use of exemptions may require written reporting. The following activities, if conducted prior to obtaining DOS and/or DOC approval, could constitute a violation of a U.S. Law:

- Talking to or emailing a foreign person - here in the U.S. or overseas - about data/technology.
- Permitting access or conducting a plant tour through sensitive areas (areas that are not necessarily classified) showing manufacturing know-how.
- Carrying technical documents (or memory device) on a business trip overseas.
- Shipping parts, components or hardware.

The key steps are to: identify what is an export, when the export may/will take place, if there are any foreign national issues and take adequate measures to obtain advance approval from the DOS or DOC. More information should be available from your organization's Export Control function.

## 7. Subcontracts baseline

Your organization is required by FAR Part 6 to provide for full and open competition when selecting subcontractors to perform work on a government contract. This competitive process is usually accomplished during the proposal phase in compliance with the your organization's procurement procedures. In limited cases, a subcontractor may be selected on a sole source basis as long as the written justification complies with the FAR. Make sure to carefully evaluate and document sole-source selections so they don't come back to bite you!

Subcontractors who are individuals, companies or quasi-government agencies must sign a subcontract with your organization to be able to perform work. Only your Subcontracts Department can sign a subcontract.

In many cases, your organization itself is the subcontractor in which case your Contracts Department will be negotiating and signing a subcontract with your prime contractor.

Your organization might decide to use a subcontractor for one or more of these reasons:

- Part of current contract team
- Incumbent staff with client
- Positive relationship with client
- To help with workload
- Unique skill set or technology
- Unique key personnel
- Extra qualifications necessary to win
- Low price reliable provider
- Strategic your organization's partner
- Fulfills a socio-economic requirement

Finalize which parts of your SOW and corresponding WBS tasks each subcontractor will perform. This is usually described at a high level in an attachment to the teaming agreement but needs to be clarified in a subcontract agreement to be compatible with the final negotiated contract requirements and project WBS.

*Subcontract Agreement.*

The subcontract agreement formally defines the work to be performed by the subcontractor and the terms and conditions for performance of that work. It's required for all subcontractors providing services. The subcontract agreement may be drafted prior to the award of the project contract but isn't finalized until after project award (for fixed price contracts subcontracts should be negotiated prior to project award). Specific items from your prime contract with the customer should flow down into your agreement with the subcontractor.

The subcontract management plan forms the basis for the technical work to be performed under the subcontract by major subcontractors and should be an attachment to the subcontract agreement. Your business unit should develop the technical aspects of the plan and a Subcontracts Administrator should develop the contractual aspects. An authorized representative, typically a Contracts Manager, will sign the agreement.

Prepare a purchase requisition in your company's online procurement system to include the subcontractor SOW. Assist the Subcontracts Administrator by reviewing the contract and the intended subcontract to ensure that all applicable Ts & Cs (which is contract slang for terms and conditions) such as contracts clauses and invoicing, have been incorporated. Note that calling them Ts & Cs will make you seem very in the know!

Support the negotiation process between your organization and the subcontractor. Ensure that the appropriate management personnel approve the intended negotiations prior to actual implementation. Make sure that all subcontract commitments and scope changes are approved by senior management and the Contracts Director.

The Subcontracts Administrator should be involved in all contractual communications with your subcontractors. Fixed price subcontracts will have some additional requirements including the need to agree on an interim payment schedule that is tied to time or specific project deliverables. The subcontractor payment schedule also needs to be synchronized with your organization's payment schedule to the government client.

*Purchase Order*

The purchase order formally defines the services and products to be delivered by a vendor and the requirements for delivery. Your organization may require a purchase order for all subcontractors providing services and products or just those companies providing short-term consulting or physical products such as computer equipment.

A request for quote may be drafted prior to the award of the project contract for costing purposes, but the purchase order isn't issued until after project award or approval of a work authorization permitting the order to be placed. The subcontract management plan forms the basis for identifying purchase order line items.

## 8. Facilities

There are numerous possible facilities configurations depending on whether your Project Staff will work at your organization's facilities or at a government site or both. Your Facilities Manager can help you define these requirements and coordinate with your client to ensure you and your staff's needs are met. Because of the criticality of these requirements, it's wise to identify and communicate these needs as soon as practical. Where possible, this planning should begin prior to contract award. Typically, but not always, on-site means working at the government's facility and off-site means residing at your organization's facilities. However, I've worked with companies that reverse these descriptions!

## 9. IT support

IT requirements are not always completely spelled out in a proposal. So, it's incumbent on the PM to carefully assess these needs both from a Project Team support perspective and/or an IT system solution delivery needs. The IT Department can assist with this assessment. Where necessary, additional or changed IT requirements may have to be factored into a revised budget and then discussed with your client through the Contracts Department.

You should establish an online contract documentation library on your organization's portal under the appropriate project site. Content in the contract repository should include standard components such as risk log, action item log and shared documents as well as files for each delivery/task order to be executed under the contract. These may be developed as needed.

## 10. Security

Some of your organization's contracts might require access to classified information. As with all support functions, there are compliance requirements that are contractually binding. Your Security Department can help you find the optimal balance between the project requirements and the Security compliance requirements. Successful customer Security inspections of your project and the organization, along with full compliance with Security requirements and minimal Security incidents ensure high customer satisfaction and repeat business.

The details of the Security requirements for your contract are most often spelled out in a DD-254 (Contract Security Classification Specification), the SOW, and Sections H, J, or some other part of the contract depending on how the client structures their RFP and contract requirements. To explain all the variations here would be nearly impossible; therefore here some points for you to ponder as you begin PgM Planning:

- Involve Security early - preferably in the capture and proposal process. It's easier to tailor a Security program and support if it's resourced and written into the proposal.
- Security can often be a direct contract charge.
- Need a safe Lock replaced? Access control system? Alarms? Classified IT system? Facility? These Security costs should be part of your PgM Plan.
- Staff clearance processing times can vary from a few days to months and are a function of the clearances required, who you hire for the team, and the customer. Your understanding of the specific customer timelines will help you plan your project schedule, budget and ability to deliver your services or products. Note that:
  - Clearances are contract specific. If a person doesn't need a clearance to work on a contract they are not authorized to have one.
  - Clearances or accesses are not always transferrable between contracts, customers, or companies in the case of a new hire.
- Facilities and IT systems needed to process classified information may have to be designed, built and approved before you can perform the work. You may be able to take existing facilities and IT systems and convert them, but this isn't always the case. The costs of setting these up may or may not be reimbursable under the contract.
- Contracts are issued to legal entities, not the organizational units. If you wish to use other organization legal entities on your project, unless they are specified in the contract, for Security purposes you must treat them like a subcontractor. These are usually handled via an Internal Work Order as the subcontracting mechanism.
- Security should work with Recruiting and HR to ensure candidate employees are screened and clearance processing starts during the on-boarding process. You should provide Security with the essential information, such as level of access required, a written justification and the contract number.
- Clearance requests are usually not submitted to a customer until a signed offer letter is on record with your organization. You can, however, initiate all the paperwork and have it ready.
- Your organization will probably not do the required background checks until it receives a signed consumer authorization release from the perspective employee. Normally an offer letter will not be issued until the company background check is completed.

## 11. Material procurement

Depending on the type of project, you may have numerous material (often spelled materiel by logistics experts) requirements that you and your Project Team will need to have ordered by a Procurement Specialist in accordance with approved government acquisitions procedures. A Procurement Specialist can walk you through the standard procurement process. It's important for you to identify all material on a bill of materials (BOM) and then carefully communicate these requirements using a purchase requisition and determine as soon as possible if any of these items fall on the project critical path. This purchase request, once approved, will become a purchase order.

A BOM is a descriptive listing of the components (hardware and software) that make up a subassembly and/or an end item product. Typically, a BOM is generated by engineering and priced by material estimating and/or purchasing based on the descriptions provided by the engineers. The BOM should be tiered from the lowest sub-assembly level to the final end item. This type of BOM is known as an "Indentured Bill of Material." This next table portrays a simple BOM for the same end item shown in indentured and non-indentured form:

| Model XYZ Computer System Non-Indentured | Model XYZ Computer System Indentured |
|---|---|
|  |  |
| Printer | Printer Subsystem |
| Cable | Printer |
| Monitor | Cable |
| Monitor cable | Monitor Subsystem |
| CPU | Monitor |
| CPU Cover | Cable |
| Hard Drive | CPU Subsystem |
|  | CPU |
|  | Cover |
|  | Hard Drive |

## 12. Stakeholder matrix

It's hard to overstate the importance of relationships to the successful management of your project. It's critical that you identify the key customer individuals who are project decision makers or influencers. Internal Departments also play a critical role in ensuring your success. Your efforts to effectively communicate with all stakeholders in a timely and responsive manner will pay huge dividends.

A project stakeholder matrix as shown in the next table is a useful management tool to document the role each stakeholder has in determining your project's success. This example illustrates a simple matrix for a small subset of project stakeholders. This matrix should include everyone you'll communicate with during the execution of your project.

| Role | Name & title | Role<br>*Key: Owner, reviewer, control, approval, input, information only* | Who in your organization is the key contact |
|---|---|---|---|
| **Your senior management** | | | |
| **Your contracts** | | | |
| **Your finance** | | | |
| **Your security** | | | |
| **Your IT** | | | |
| **Client PM** | | | |
| **Client technical** | | | |
| **Client contracts** | | | |
| **Client finance** | | | |
| **Client operations** | | | |

## 13. Configuration change control

As the PM, you should implement the process for measuring, reporting and controlling changes to the project requirements as well as processes to be used in assessing the impact of requirements changes on product scope and quality, and the impacts of requirements changes on project schedule, budget, resources and risk factors. This includes:

- Measure the progress of work completed at major and minor project milestones.
- Compare actual progress to planned progress and implement corrective action when actual progress doesn't conform to planned progress.
- Specify the methods and tools that will be used to measure and control schedule progress.
- Identify the objective criteria that will be used to measure the scope and quality of work completed at each milestone, and to assess the achievement of each schedule milestone.
- Specify the methods and tools that will be used to track the project cost. Identify the schedule milestones and objective indicators that will be used to assess the scope and quality of the work completed at those milestones.
- Indicate whether earned value tracking will be used to report the budget and schedule plan, schedule progress, and the cost of work completed.

## 14. Quality assurance and control

Standardization has become one of the best ways to ensure a greater level of quality throughout all areas of management, from processes all the way through to the management of projects. ISO Standards are therefore a vital tool in the PM's toolbox, ensuring a better level of project success through universal systems that are designed to set clear guidelines for projects, from design all the way through to implementation and review. Your project should comply with the general guidelines of ISO9001 as illustrated in the next figure:

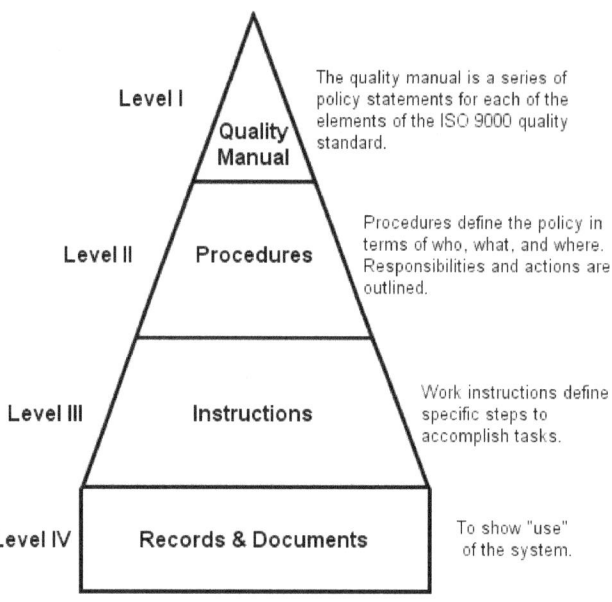

The quality manual is a series of policy statements for each of the elements of the ISO 9000 quality standard.

Procedures define the policy in terms of who, what, and where. Responsibilities and actions are outlined.

Work instructions define specific steps to accomplish tasks.

To show "use" of the system.

At the other end of the spectrum, your organization may manage smaller services and development projects for years on end with a less formal concept of quality management. However, as a principle, every project should be managed with an appropriate and tailored application of quality management approaches and methodologies, and with a commitment to and understanding that continual improvement in project performance through proactive monitoring, measurement and correction is the standard approach at your organization.

### A. Summary of key aspects of ISO 9001:2008

- The quality policy is a formal statement from management, closely linked to the business and marketing plan and to customer needs.
- The quality policy is understood and followed at all levels and by all employees. Each employee works towards measurable objectives.
- The business makes decisions about the quality system based on recorded data.
- The quality system is regularly audited and evaluated for conformance and effectiveness.
- Records show how and where raw materials and products were processed to allow products and problems to be traced to the source.
- The business determines customer requirements.
- The business has created systems for communicating with customers about product information, inquiries, contracts, orders, feedback, and complaints.

- When developing new products, the business plans the stages of development, with appropriate testing at each stage. It tests and documents whether the product meets design requirements, regulatory requirements, and user needs.
- The business regularly reviews performance through internal audits and meetings. The business determines whether the quality system is working and what improvements can be made. It has a documented procedure for internal audits.
- The business deals with past problems and potential problems. It keeps records of these activities and the resulting decisions and monitors their effectiveness.
- The business has documented procedures for dealing with actual and potential non-conformances (problems involving suppliers, customers, or internal problems).
- The business:
  - Makes sure no one uses a bad product.
  - Determines what to do with a bad product.
  - Deals with the root cause of problems.
  - Keeps records to use as a tool to improve the system.

As part of the PgM Planning process, you should prepare a Quality Assurance/Management Plan to regularly evaluate overall performance via the identification and definition of both quantitative and qualitative quality indicators which, if properly applied, will provide confidence that the end product will meet the customers' needs and expectations. The quality assurance plan can be a stand-alone document, or, in some cases, a section of the PgM Plan.

In general, the government's quality assurance requirements are that:

- The contractor shall have a QA system that is approved by the Government or customer.
- The contractor shall use this QA system on any products produced and sold to the Government or customer.
- The contractor shall have objective proof that the product meets all of its product requirements.
- The contractor shall assure that subcontractors also meet the above requirements.

Quality control should be emphasized to the lowest possible level on your project. The PM, QC Representative and Work Group Leaders, etc., should all share in the overall responsibility to ensure high quality.

*B. Roles and Responsibilities*

The Corporate QC Manager is responsible for your organization's QC Project. In turn, the QC Manager will assign specific responsibilities to certain individuals within the QC chain in your organization. It's incumbent on you to ensure that all the designated responsibilities are completed and adhered to at all times. The following QC positions, along with respective responsibilities, form the QC for your organization. Any reassignment of the responsibilities listed below must be approved by the QC Manager. Of course, if there isn't a QC Manager, you might have to assume that responsibility.

The QC Manager has the following responsibilities:

- Establish, implement, monitor and maintain the QC program.
- Review and approve revisions made to the QC Program.
- Review program adherence and inspect delivery of contracted services.
- Perform daily checks of all areas using checklists, periodic inspection and direct observation.
- These checks should be documented and maintained on-site for review.

The Quality Control Specialist, or alternate, should:

- Update checklists to be contract specific.
- Coordinate with the Contracting Officer or the COTR.
- Assign individuals to be responsible for the actual inspection and corresponding remarks.
- Make revisions to the final checklist.
- Perform on-site QC inspections.
- Provide corrective action to all discrepancies.
- Establish frequencies for performing inspections.
- Provide a copy of all inspection reports and corrective actions to the Corporate Quality Control Manager.
- Implement a Proactive Quality Control Program.
- Provide employee training on the processes and procedures of the quality control program through on-the-job training and staff meetings.
- Empower the Project Team to be proactive by involving every employee in the process.

Task Leaders should check all completed work for accuracy, timeliness, completeness and overall quality.

Individual employees should double-check their work to ensure high quality and compliance with contract requirements.:

## 15. Risk management

*The basic objective of risk management is to minimize the number of risks that become issues.*

Having participated in over a hundred reviews of government projects of all sizes and complexity, I can sadly report that only a handful had implemented even a somewhat satisfactory degree of risk management and control. Most PMs didn't even give risk lip-service! And few RFPs required it. Hopefully, you'll take this section to heart and will have mitigated most of the major risks on your project before they occur and have fewer major issues to deal with during your project's lifecycle.

Project risk management is the systematic process of identifying, analyzing, and responding to project risk. Project risk is an uncertain condition or event that, if it occurs, has a positive or negative effect on at least one project objective. A risk may have one or more causes, or triggers, with one or more impacts. Risk conditions may include aspects of your organization's project management environment, contractual obligations, dependencies or stakeholder relations.

Risks can include the possibility of suffering a negative impact to the project, whether it's decreased quality, increased cost, schedule slip or project failure. The process should assign specific responsibilities for the management of risk and prescribes the documenting, monitoring and reporting processes to be followed.

This section presents the process for implementing proactive risk management. Risk management is a project management tool to assess and mitigate events that might adversely impact the program. Successful implementation of risk management will increase the program's likelihood of success. This process will:

- Serve as a basis for identifying alternatives to achieve cost, schedule and performance goals
- Improve service and product quality

- Assist in making decisions on budget and funding priorities
- Provide risk information for Progress Reviews or milestone decisions
- Allow monitoring the health of the project as it proceeds through the lifecycle

The five key elements of risk management are:

A. Risk management planning
B. Risk identification
C. Risk analysis (qualitative and quantitative)
D. Risk response planning
E. Risk monitoring and control

This figure illustrates the risk management process as it relates to a more complex software development project lifecycle:

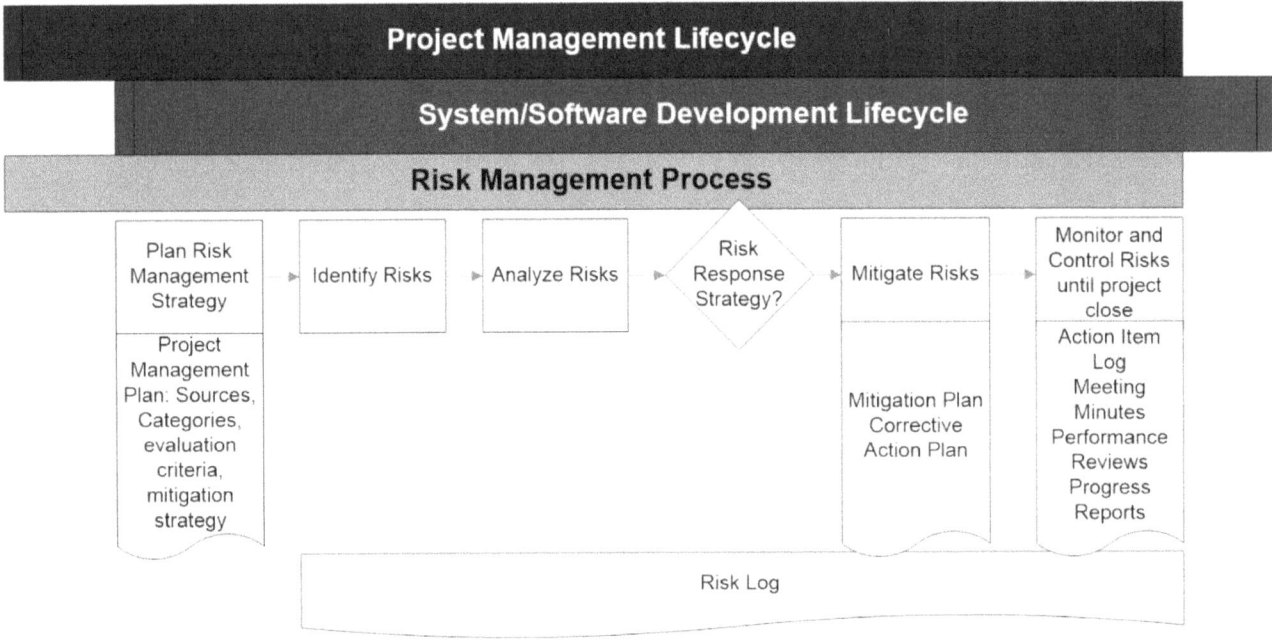

Something in writing should define the roles and responsibilities and processes to be used by project key players to manage and control risk. Also, get agreement on acceptable risk tolerances for those times when will you need to sound the alarm! For instance, when there is more than a one-week schedule slip or a projected cost overrun will occur by a specified date.

*Risk identification*

Risks should be identified at the beginning of every project/task order and should include:

- Business (cost and contract)
- Technical
- Schedule
- Resource availability
- Programmatic

It doesn't have to take very long to identify "what if scenarios" but you should definitely involve the key Project Team members. The product of this exercise is several "what-if" scenarios such as:

- What if key personnel leave?
- What if requirements change?
- What if your budget is reduced by 20%?

Some of the techniques that you can use to identify and document the various risks that might occur on your project are:

- Brainstorming
- DELPHI technique
- Expert interviewing
- Historical records
- Checklist

Some of the common situations that can negatively impact the project schedule are:

- Tasks on the critical path
- Overly optimistic estimates in terms of cost, schedule, technical or some combination of the three
- Tasks reliant on external dependencies
- Tasks with a large number of predecessors
- Changing specs/requirements

Some of the common situations that can negatively impact the project resources are:

- Scarce resources
- Many different resources to coordinate
- Changes in corporate funding priorities
- Changes in costs
- Unqualified resources

Some of the common situations that can negatively impact the project scope are:

- New products
- Changing or creeping requirements
- Availability of technology
- Unexpected defects
- Supplier failure

*Risk analysis*

The key to risk analysis is to isolate the risks that have the highest probability of occurring and the greatest potential negative impact of the project. Then to identify steps that can be taken to minimize the probability of these risks happening or at least minimize the negative impact if they do occur. The following diagram illustrates a technique to identify the potential cause of a specific project risk. Of course, you can also use this technique to identify the cause of a problem that has already occurred.

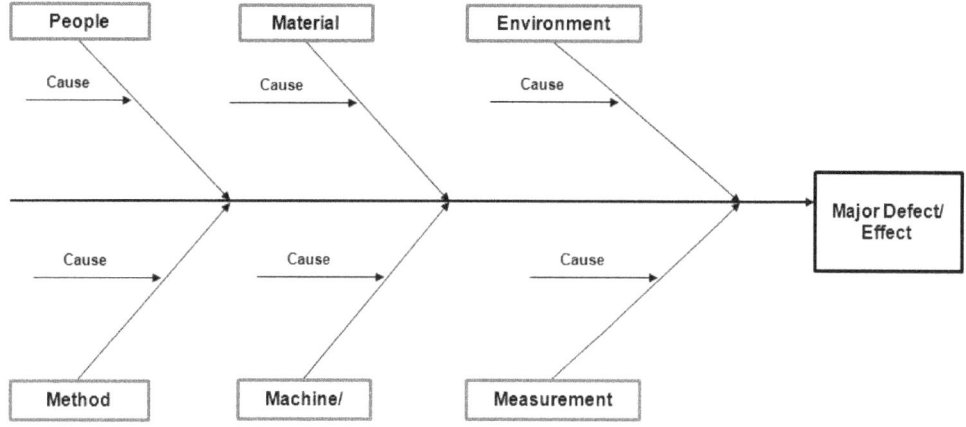

*Qualification risk analysis*

There are three key risk analysis components – the potential risk events, the probability that the risk will occur and the potential impact on the project if the risk occurs. The following matrix can be used to evaluate a potential risk:

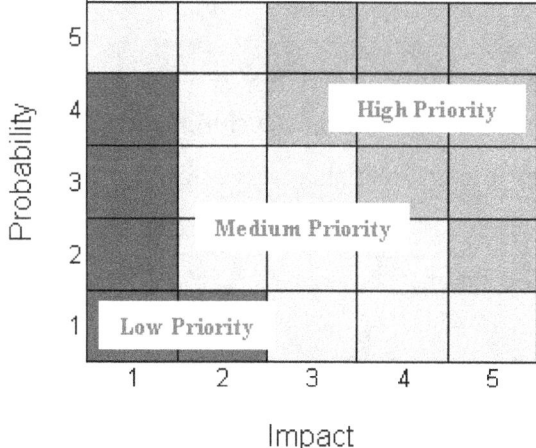

Risks that fall into the medium-shaded cells (the upper right) of the matrix are the highest priority and should receive the majority of risk management resources during response planning and risk monitoring and control. Risks that fall into the light-shaded cells (the middle) of the matrix are cautionary risks, followed by the lowest risks that fall into the darkest-shaded areas (the lower left).

The following two charts illustrate a standard set of risk impact probability and impact definitions. During risk analysis, the potential impact of a given risk is assessed and an appropriate risk probability level is selected.

| Likelihood (Probability of risk happening) | |
|---|---|
| 1 | Not Likely |
| 2 | Low Likelihood |
| 3 | Likely |
| 4 | Highly Likely |
| 5 | Near Certainty |

Risk probability definitions

| Consequence (Impact to the project if risk happens) | |
|---|---|
| 1 | Minimal |
| 2 | Some |
| 3 | Moderate |
| 4 | Significant/High |
| 5 | Critical/ Severe |

Risk probability impact

*Quantitative risk analysis*

There are several major quantitative risk analysis techniques that may be used on more complex projects. Some cost estimating techniques for risk analysis, for instance, include:

- Analogous Estimating (Top-down)
- Bottom-up Estimating
- Parametric
- Three-Point Estimating
- Project Management Software
- Vendor Bid Analysis
- Reserve Analysis (Manager's Reserves)
- Interviewing
- Decision tree
- Simulation (Monte Carlo)

*Risk response (mitigation) planning*

There are four basic strategies for treating (mitigating) the risk on your project.

1. Avoidance. This is using an alternate approach that doesn't have the risk. This mode isn't always an option, but this is the most effective risk management technique if it can be applied.

2. Control: The Risk Management Guide from the Defense Systems Management College (part of the Defense Acquisition College -- dau.edu) defines this mode as "the development of a risk reduction plan and then tracking to the plan." The key aspect is the planning by experienced persons. But the plan itself may involve parallel development programs.

3. Assumption: This is simply accepting the risk and proceeding without any mitigation.

4. Risk Transfer. This is attempting to pass the risk to another program element. This could entail passing the risk onto a subcontractor, customer, or another outside your organization. My worst experiences with this are when a PM passes a poor performing staff member to a different Department in the organization instead of dealing directly with the performance issues.

This risk register (risk log) should be used to document project risks and mitigation steps for the high and, where practical, medium probability and impact risks as shown here:

| Risk ID Number | Threat Event | Probability | Impact | Mitigation Strategy | Follow-up Date | Status | Responsible Team Member or Manager |
|---|---|---|---|---|---|---|---|
|  |  |  |  |  |  |  |  |
|  |  |  |  |  |  |  |  |
|  |  |  |  |  |  |  |  |
|  |  |  |  |  |  |  |  |
|  |  |  |  |  |  |  |  |

Once you have identified a potential risk that has a potential impact and probability of occurrence that is great enough, you'll need to select the best risk mitigation approach.

A matrix similar to this can be used to analyze each mitigation alternative:

| Criteria | Alternative 1 | Alternative 2 | Alternative 3 |
|---|---|---|---|
| Cost impact | | | |
| Schedule Impact Low/Med/High | | | |
| Risk of Failure Low/Med/High | | | |
| Resource Requirements | | | |
| Total score | | | |

If there is no way to eliminate a specific risk within reasonable time and dollar constraints, look for how to minimize the probability or consequence of the risk. Some of the techniques that can be used to mitigate some of the more project common risks are:

- Look for how to make the risk go away so that is it no longer a risk.
- Develop a workaround plan.
- Implement corrective actions.
- Process a change request.
- Update the project and risk response plans.
- Document lessons learned.

If risk elimination or minimization is too difficult or impossible and all you can do is to wait and see if it occurs, then you might devote some time to developing a plan to be ready to react to the risk.

The following list describes many of the risk scenarios that can occur on government projects.

- Poor Start-up, late established baseline
- Viable risk methods proposed and available but seldom used
- Cost based on proposed productivity but productivity isn't measured during life of project
- PM lacking in prior, successful experience, i.e., never completed a successful PM job before or never managed a fixed price program
- Initial and follow-up planning insufficient to provide clarity and completeness

- Little up-front attention paid to necessary architecture of the intended system and, when attention is paid, no viable analyses of "workability/do-ability" of the architecture is defined
- Planning and progress tracking mostly by subjective means vs. objective measures
- CM/CC methods/controls insufficient to meet evolving system
- Risks not identified early-on and/or no risk identification-mitigation program in effect
- Policy adherence given "slow-roll" by all concerned
- Employees leaving early to seek "better opportunities"
- Metrics not followed
- Management reviews seldom held
- During reviews, "accomplishments" presented without being tied to things that were "planned"
- Acceptance criteria for deliverables not sufficiently defined/agreed upon with the customer prior to delivery or customer acceptance of criteria not documented
- Requirements definition(s) not clearly nor completely specified sufficient for the viable use of the intended implementers
- System's performance characteristics not adequately identified, allocated, tracked, nor well understood
- Efforts in progress not consistent with baselined planning
- Capabilities continually moved to later releases to "ease" schedule pressures
- Sufficient QA methods not employed
- Project Team not committed to cost/schedule objectives because they view superior technical performance more important
- A different team is performing the project than the one that proposed it and the technical approach is now deviating from the original vision
- PM doesn't believe in assigning formal action items at staff meetings
- The Bill of Material is inadequately defined or controlled
- Viable task interdependencies not defined/maintained and/or used as a management "tool"
- No viable/approved "business case/plan" defined or used to justify the project
- "Completion criteria" not defined for individual task assignments
- Adequate training/mentoring of management staff not done
- The PM has never fully read the contract (amazing how many times I've discovered this to be the case!)
- PM doesn't set and communicate priorities between cost, schedule and superior technical performance

- Time to coordinate with teammates or vendors not factored into the schedule
- GFE not available
- PM doesn't effectively delegate and becomes a single-point of failure, delaying all decisions
- The PM doesn't develop nor implement a communications plan

## 16. Action planning

A project list of actions is probably the single biggest risk mitigation step you can take. These actions can tracked using one of the many available online tools.

Your action plan should identify the major actions that need to be taken to successfully compete your project and who will be responsible and a due date. These actions may be derived from the steps described in this section.

My consulting clients have had great success using a yellow/green/red color scheme to prioritize actions. This ensures management attention is given to the most critical and important actions.

Coordinate this action plan with all internal and external interfacing groups. Maintain this action list through the life of the project and use it during your project status meetings.

## 17. Kick-off meetings

It's important to have well thought out, proactive project kick-off meetings. This is similar to making sure a construction contractor team understands the blueprints before laying a new building foundation. All company stakeholders should be invited to internal meetings and all client stakeholders should be invited to client kick-off meetings unless otherwise specified in the contract or by the client.

See Appendix B for a Project Start-up Checklist.

The degree of preparation for a kick-off meeting will quite often determine its successful outcome. Before the kick-off meeting, you should:

- Determine the purpose and objectives of the meeting.
- Acquire meeting facilities.
- Set the ground rules for participation and discussion.
- Invite the attendees based on a need to know, 5 to 10 days in advance, if possible.
- Develop and provide the meeting agenda to the attendees before the meeting.
- Develop any visual aids, notes or handouts.
- Assign someone to take minutes and record actions.

In particular, remember that you'll get much more enthusiastic buy-in from meeting attendees by emailing a detailed agenda and soliciting input at least one day before.

Every agenda topic should have one of three objectives: To pass on information, to come to a decision or to gather information. Otherwise, consider eliminating it!

Nothing can derail a new project more quickly than a poorly run kick-off meeting. Pass out an agenda in advance and begin and end on time. Make sure you stick to the agenda. Try to draw people out. Don't assume silence is consent. Finally, be sure to record decisions and action assignments, and then follow up on action status.

The key elements of a successful internal Project Team kick-off meeting are:

- Define project success criteria so the team understands your client's expectations.
- Identify any risks, challenges and project constraints. Take the time to respond to everyone's concerns about project success. Don't assume silence is concurrence. Now is the time to encourage open discussion while there is still time to change the plan.
- Make sure everyone understands the necessary project control and status requirements including documentation standards and quality reviews. Also, clarify time-keeping and invoicing requirements for other direct charges like travel expenses.
- Go over the tools, documents and support needed from the client.
- Record decisions and action assignments.

A strong beginning will greatly increase the probability of project success. And, it will hopefully keep you out of trouble!

Typical internal kick-off meeting agenda items include:

- Introductions
- Scope and objectives
- Risks, challenges, and project constraints
- Project approach
- Success factors
- Team members and project organization chart
- Roles and responsibilities
- Expected client involvement
- Work plan review
- Major milestones and deliverables
- Timeline
- Training schedule
- Project control techniques
- Quality review
- Documentation standards and storage
- Project status reporting
- Travel plans
- Invoicing requirements for ODCs
- Logistics for the team
- Tools needed from the client
- Questions needing immediate answers
- Consensus on Start-up activities
- Preliminary documentation requests
- Risk review
- Action review and next steps
- Security

After the meeting, make sure to distribute meeting minutes. This should include an action item list that identifies the responsible party and suspense date for each action. The date of the next meeting should also be shared. Follow up on action item status on a regular basis.

These guidelines also apply to the client kick-off meeting with the addition of any requirements specified in the contract. Coordinate with your government counterpart to try to get all the key client stakeholders from deciders to influencers to attend and conduct the meeting as soon after contract award as possible.

Probably the most important objective of client kick-off meeting is to achieve a clear agreement and understanding of the key work elements, deliverables and schedule as well as what the government is contractually required to do.

## 18. Project management (PgM) plan

There are several reasons PMs give for why they don't think they need a PgM Plan including:

- Good project management is "nice to have" but not a necessity.
- My projects are all crises; I don't have enough time to plan.
- Structured project management is only for large projects.
- My projects require creativity and can't be predicted with any certainty.

None of these are valid reasons for not having a well thought out PgM Plan.

Many major government projects require a PgM Plan as a contract deliverable. This document describes how you'll manage the project to successful completion. Your plan should include the following:

- Project charter (roles and responsibilities)
- Project management approach or strategy
- Scope Statement
  - Project objectives
  - Project deliverables
- Work Breakdown Structure (WBS)
- Cost estimates, schedule and responsibility assignments for deliverables

- Measurement baselines for scope, schedule and cost

- Major milestones, reviews and target dates

- Required staff

- Quality assurance and control

- Configuration management and control

- Risk management

- IT, facilities and Security plans

- Action plans

The PgM Plan should describe the configuration management, project metrics and quality assurance process to be employed on the project in addition to any project management activities. A description of these processes will be covered in more detail in the Project Execution, Monitoring & Control section. There is also a comprehensive PgM Plan template in Appendix A.

The PgM Plan is a living document. It also serves as a valuable transition document if a new PM comes on-board. As such, you should regularly revisit and update it to reflect current project status and to make sure you haven't neglected an important aspect of the plan. Remember to update the revision history section as well as any headers/footers in order to maintain document and version control. When practical, a draft of these plans should be prepared prior to contract award to ensure proper pre-award planning takes place.

# Section III: Project execution, monitoring and control

The project execution, monitoring and control phase typically has the longest duration. It's the phase during which project activities are carried out and deliverables are constructed and presented to the customer for acceptance. To ensure that the customer's requirements are met, the PM monitors and controls the necessary activities and resources. A number of management processes are undertaken to ensure that the project proceeds as planned. These are discussed in this section.

This project execution, monitoring and control section is comprised of the following eleven activities:

1. Project execution (implementation)
2. Financial management
3. Time (schedule) management
4. Change management
5. Quality management
6. Risk monitoring & control
7. Issue management
8. Subcontract and procurement management
9. Human resource management
10. Communications management
11. Customer relationship management & contract growth
12. Security support management

The following table summarizes the key information contained in this section:

| Tasks | Deliverables (Outputs) | References/ Tools | Interfaces | Key Points |
|---|---|---|---|---|
| . Project execution (implementation) | -Project deliverables -PgM Plan updates -Activities #2 - #10 which are designed to deal with the six primary project constraints: Scope, cost, schedule, human resources, quality & risk | -Project Start-up checklist – Appendix B -Corporate project execution review slides | Contracts, project control, billing, subcontracts, Security, HR, facilities, IT | -Your project should: -Meet or exceed project mission and objectives -Result in a satisfied client -On or ahead of the agreed upon schedule and on or under budget -Result in winning the client's trust -Project staff is motivated at completion to work on another project -The client sells your services to other clients for you -Result in follow-on business -Keep track of project metrics -Continue to keep all key stakeholders "in the loop" |
| 2. Financial management | -Project cost summary -ETC, EAC, etc. -Project status reports -Financial status -Forecasts -Re-planning and revisions to the baseline -Internal and customer reports -Financial aspects of project reviews | -Financial System -PM Dashboard -Job status reports | -Pricing, program controller, finance analyst/biller, contracts, subcontracts | -Understand labor rate composition -Review and validate monthly financials   - Generate and maintain monthly forecasting template   - Generate EACs for FFP contracts   - Participate in budgeting process   - IFR review and/or initiation   - Accrual tracking and estimating   - Procurement requisitions   - Time phased budget management plan |
| 3. Time (Schedule) management | -Milestone charts, activity lists, Gantt charts, activity networks, critical path networks, PERT charts -Timesheet register | -Automated scheduling tool like Microsoft Project | | -Understand critical path -Gather status on a regular basis -Be proactive, develop corrective action plans & communicate them to your client |
| 4. Change management | -Change process defined in PgM Plan -CMM tool for change control -Cost accounting processes including EVM | -Change control tool like Caliber RM or Borland Star Team Enterprise | -Quality Director | -Consider having a Change Control Board (CCB) for more complex projects |
| 5. Quality management | -Quality process defined in PgM Plan | | -Quality Director | -Specify use of quality control processes including conformance to work processes, verification & validation, joint reviews, audits & process assessment |

| | Tasks | Deliverables (Outputs) | References/ Tools | Interfaces | Key Points |
|---|---|---|---|---|---|
| 6. | Risk monitoring & control | -Risk process defined in PgM Plan<br>-Risk register (log) | | | -Assess business, technical, schedule, resource and programmatic risks for impact & probability<br>-Maintain a risk register (log) |
| 7. | Issue management | -Project action item list | | -All key stakeholders | -Maintain and regularly status action item list |
| 8. | Subcontract & procurement management | -Bill of Materials<br>-Purchase order | -Financial System purchase request (PR) | -Procurement<br>-Proposal manager<br>-Contracts | -Work closely with Procurement Specialist<br>-Identify everything on bill of materials<br>-Anticipate time to obtain PR approval |
| 9. | Human resources management | -Recruiting requisition<br>-Performance reviews | | -Al Departments | -Conduct on-time personnel reviews<br>-Recognize achievements<br>-Keep written record of employee discussions, warnings, etc. |
| 10. | Communica-tions | -Job status report (internal & client)<br>-Project All Hands Meetings<br>-Project Newsletter<br>-Weekly Staff Meetings<br>-Technical Interchange Meetings<br>-Design Reviews<br>-Test Readiness Reviews<br>-Customer Reviews<br>-Social Events<br>-Email<br>-Teleconference<br>-Contracts Letter | | | -Report any significant impacts to the project<br>-Really listen to project team members to understand any risks and issues<br>-Don't waste time with meetings that lack a clear objective<br>-Distribute information in a timely fashion |
| 11. | Customer relationship management & contract growth | | -Your organization's Busing Development Guide | -Business development organization | -Know your stakeholders<br>-Understand your contract<br>-Be careful what your promise<br>-Manage expectations<br>-Don't let your scope creep<br>-Don't take things for granted<br>-Don't surprise your client<br>-Be proactive / offer solutions<br>-CRM guidelines<br>-Grow your contract |
| 12. | Security resource management | -Plan for ensuring Security require-ments are met | | -Group Security Team<br>-Customer Security | -Ensure all Security requirements can be met and are coordinated as appropriate with the customer<br>-Timeline planning for clearances, facilities, IT systems can be critical path items |

## 1. Project execution (implementation)

Project execution success means that your project:

- Meets or exceeds project mission and objectives
- Results in a satisfied client
- Is on or ahead of the agreed upon schedule and on or under budget
- Results in winning the client's trust
- Project staff is motivated at completion to work on another project (with you!)
- The client sells your services or products to other clients for you
- Results in follow-on business

During project execution, you distribute work packages and execute key management areas based on areas defined in the PgM Plan. Task leads break down the information into task lists and execute those tasks. As the PM, you're primarily responsible for executing the project in the most effective way that deals with these project constraints:

- Scope
- Cost
- Schedule
- Human resources
- Quality
- Risk

System development projects, as opposed to labor support contracts, have a unique set of project execution phases depending on the type and complexity of the project.

While the Project Team is busy developing project deliverables, you'll need to implement a series of management processes to monitor and control the activities being undertaken by the Project Team. Monitoring the work involves reviewing technical, cost and schedule progress against the baseline plan.

The monitoring and controlling process consists of activities performed to observe project execution so that problems are identified in a timely manner. This ensures ample time to correct any issues found to control project performance. Project performance is observed and measured regularly to identify project variances from the approved plans documented in the PgM Planning process.

The monitoring and controlling process also involves controlling changes and prioritizing and recommending mitigation strategies for risks anticipated on the project. At a high level, you should monitor ongoing project activities against the approved plans and project performance baselines. You should also ensure only approved scope, cost, schedule, quality, and personnel changes are implemented.

It's important to identify the project success factors to be measured at the beginning of the project execution phase. However, keep in mind the available resources to track these metrics on a regular basis. Avoid falling into the trap of too much oversight and too little performance! It's a delicate balance.

The key roles for the monitor and control steps as they relate to project execution are summarized in the following table and discussed in the following subsections:

| Key Roles | Project Execution Responsibilities |
|---|---|
| PM | <ul><li>Financial System setup – carefully proof for errors and correct immediately</li><li>Understand labor rate composition</li><li>Review and validate monthly financials</li><li>Generate and maintain monthly forecasting template</li><li>Generate Estimates at Completion (EAC) for FFP contracts</li><li>Participate in budgeting process</li><li>Internal Finance Review (IFR) review and/or initiation</li><li>Accrual tracking and estimating</li><li>Procurement requisitions</li><li>Time-phased budget management plan</li></ul> |
| Project Control Analyst (or PM if no Project Control Analyst is assigned) | <ul><li>Financial project status data</li><li>Forecasts</li><li>Re-planning and revisions to the baseline</li><li>Internal and customer reports</li><li>Financial aspects of project reviews</li></ul> |
| Program Controller | <ul><li>Program financial analysis and decision support</li><li>Program financial reporting</li><li>Budgeting and forecasting maintenance and support</li><li>Program Level adherence to corporate policies and procedures</li><li>Program financial training and mentoring</li><li>EAC Review for FFP contracts</li><li>Pricing/estimation support</li><li>Accrual tracking/estimating</li><li>Procurement Requisitions (PR)</li></ul> |
| Billing Administrator | <ul><li>Contractual set-up in the Financial System</li><li>Prepare and submit invoices</li><li>Resolve invoice problems</li></ul> |

## 2. Financial management

Part of your PM job is to ensure that the project is completed within the allocated and approved budget. Budget management is concerned with all costs associated with the project, including the cost of human resources, equipment, travel, materials and supplies. Increased costs of materials, supplies, and human resources, therefore, have a direct impact on the budget. Just as task duration estimates are tracked carefully against actual costs, the actual costs must be tracked against estimates.

Changes to the scope of the project will most often have a direct impact on the budget. Just as technical scope changes need to be controlled and managed, so do changes to the project budget.

It's the responsibility of the PM to closely monitor the financial performance of the project and take responsibility for addressing cost-related issues as they arise. In addition, you should always be aware of the effect your decisions may have on the total cost of the project, both before and after the product development or service is implemented.

*Roles and responsibilities*

As the PM, you're ultimately responsible for ensuring the financial baseline is established and for reviewing and validating monthly financial results. Your primary responsibilities, depending on the type and complexity of your project and how your organization functions, are:

- If not already done, set up the project in the Financial System and carefully proof for errors and immediately correct them.
- Understand labor rate composition.
- Review and validate monthly financials.
- Generate and maintain monthly forecasting template.
- Generate EAC for FFP contracts.
- Participate in the budgeting process.
- Review accrual tracking and estimating.
- Prepare procurement requisitions.
- Develop a time phased budget.

Depending on the size and scope of your project, you may have a Project Control Analyst who plays a major role in monitoring progress by supporting you with the business aspects of the project such as establishing the WBS, opening job numbers and establishing the budget baseline. This support might have been included in the budget estimate or might be an overhead expense. Otherwise, you'll be responsible for fulfilling this role. It includes the following tasks:

- Financial project status data
- Forecasts
- Re-planning and revisions to the baseline
- Internal and customer reports
- Financial aspects of project reviews

*Cost accounting tasks*

There are several cost-accounting tasks that you'll be responsible for either initiating or supporting during the execution of your project.

- Project overview
  - Your organizational structure
  - Labor rates
  - Accounting org chart
  - Accounting calendar
  - Department codes
  - Indirect Departments
- Time and Labor
  - Charge codes
  - Approve timesheets
  - Approve labor adjustments
  - Ensure employees compliance
  - Labor qualifications forthcoming
  - Subcontractor timesheet tracking (if applicable)
- Expense Reports
  - Approve expense reports/cash advances/travel authorizations
  - What is allowable/reimbursable on your contract?

- Purchasing
  - Approve Requisitions
  - Receive materials
  - Approve internal financial reports
- Project Summary Reports
  - Closing notifications
  - Accruals
  - Burn rates
  - Approve job status reports
- Tools
  - Project management dashboard (if one exists)

As you can see, there are quite a few terms and activities that impact your ability to successfully monitor and control the financial aspects of your project. Your Finance Department can assist you with each of these tasks. Don't be reluctant to ask for assistance!

*Cost components*

As PM, you'll need to understand the components that make up your labor and indirect costs (see the next figure). In my experience, too few project managers are given the necessary training or opportunity to fully understand these cost elements and the impact they have on project success. People in the Finance Department, usually over-worked, don't have the time or patience to help. So, it's up to the PM to make sure you grasp these concepts!

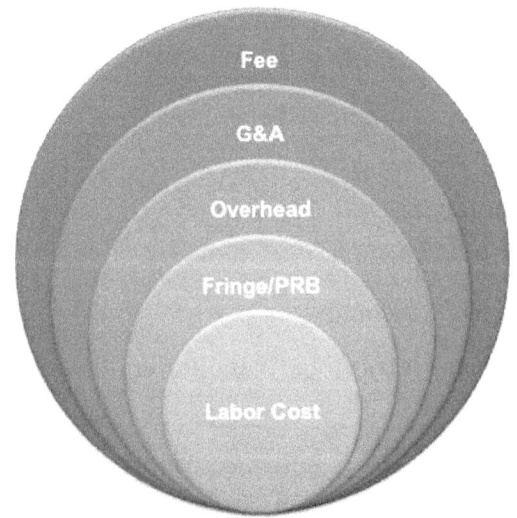

Different indirect rates, comprised of the factors shown above, will be applied to the labor charges on your project. This information is very company proprietary and competition sensitive as are project status reports. All company overhead, G&A and other rate information must be protected either by shredding or placement in a locked file. This includes reports that reflect vendor and subcontractor billing rates. Your organizations Controller can explain these cost components and how they impact your project financial baseline.

*Charge numbers*

As a PM, you should understand how the company assigns charge numbers. So, you should familiarize yourself with the following terms (details are generic examples only):

- Bus Units or groups represent legal entities, and your organization may be a group of entities.
- An account is directly in support of a contract. Requires a Project ID, Activity ID. Example: office supplies & business meals. Others won't require a Project ID but will have a Department ID and the Burden will be reimbursable to the employee but the government will not reimburse you. Examples are charitable donations, softball team, holiday party.
- Each Burden ties back to a Group, Operating Unit and Department ID.
- Each project ID is identified with a specific customer contract. You can have many projects to one contract, but the contracts tie back to the legal entity that was awarded the contract.
- Each activity ID is identified with a project #. You can have many activity IDs for a project.
- Accounts may be further broken down into individual IDs for specific labor categories, indirect and direct charges.

As an example, assume your organization has a Nine (9)-digit account number. The 1st four numbers might represent the indirect or contract (direct) charge number. The middle three could denote the task level or Department number. The final two might be the specific contract suffix numbers to spread costs to end-use.

- Balance sheet accounts are 0100-0399
- Indirect Account numbers are 0400 -0599
- Fringe accounts numbers are 0600-0699
- G/A accounts are 0700-0799
- Unallowables are 0800-0899 and allocation accounts are 0900-0999

Every organization has a financial accounting system. Cost data to support billing is accumulated within the project module, the official cost data source. The project module accumulates the cost source data from the general ledger, accounts payable and time and labor feeder systems. This contract data is subject to rigorous validation, and review and approval controls at the point of entry into the system. On a monthly basis this data should be reviewed again by someone on your Project Team to ensure accuracy of the cost data. Ultimately, the PM is responsible for the accuracy of the project data in the financial system.

*Earned value management (EVM)*

Earned value has proven to be an effective analytical methodology to identify project cost and schedule status, especially on more complex projects. However, it's important to make sure to incorporate the necessary resources and schedule time to accommodate EVM during the proposal and pre-execution phase. Because of the overhead labor required to correctly implement earned value, it's not applicable to many types of contracts. Unfortunately, I've seen it mis-used by customers to micro-manage services contracts.

EVM is a schedule and cost system used to assist Project Teams and customers in assessing progress, provide early warning for corrective action and estimates of future schedule and cost based on current progress. This is particularly effective at keeping overly-optimistic software developers honest who seem to always be 90% complete but never actually done!

Cost control metrics commonly used, at a minimum, to measure progress are:

- Cost variance
- Cost performance index
- Cost vs. funding
- ODC costs vs. plan
- Remaining budget vs. budget at completion
- Subcontractor cost variance

The next table is an example of a spreadsheet used to record planned vs. actual $ costs on a contract:

| | 14 May | 21 May | 28 May | 4 Jun | 11 Jun | 18 Jun |
|---|---|---|---|---|---|---|
| **Actual Staff By Month** | 28.9 | 35.5 | 36.3 | 45.4 | 34.8 | 42.3 |
| **Actual Hours By Month** | 1,271 | 1,279 | 1,597 | 1,227 | 1,530 | 1,523 |
| **Actual Hours Cumulative** | 23,980 | 25,259 | 26,856 | 28,083 | 29,613 | 31,136 |
| **Planned Hours By Month** | 1,034 | 984 | 1,255 | 888 | 1,611 | 1,318 |
| **Planned Hours Cumulative** | 29,427 | 30,411 | 31,666 | 32,554 | 34,165 | 35,483 |

This topic is complex. More specific guidance, if EVM is applicable to your project, can be found in several readily available earned value management training courses.

*Financial approval process*

The financial approval process has several steps that must be followed to ensure that your organization continues to comply with government regulations. See the following cost accounting process figure for a VERY general process overview. PMs should make sure they are very familiar with this process.

Prepare Contract Set-up in finance system (one day)

↓

Biller sets up contract in accounting module & coordinates activity names with PM (one day)

↓

Time Administration & email sent to Project Team members

↓

Month end: Project Status Report (PSR) run & emailed to PM for approval (in writing); Action per Sarbanes Oxley action taken (one day)

↓

PSR Not Approved – PM refers problems to Biller for resolution. Corrected PSR re-run and re-emailed to PM for approval (one day)

↓

PSR Approved – Invoice delivered to client for payment

There are three key players in the management of the cash flow process: you as the PM, the Billing Administrator assigned to your contract and the Contracts Manager. The goal is to recover all contract costs through billing. Contract costs are all direct costs specific to a contract plus indirect cost allocations.

## 3. Time (schedule) management

Schedule or time management is the process of recording and controlling time spent by staff on the project. As time is a scarce resource within projects, each team member should record time spent undertaking project activities and the appropriate Financial System charge number(s). This will enable you to control the amount of time spent undertaking each activity within the project.

Schedule status should be collected and/or reported weekly or monthly. Weekly status collection is best for larger complex projects where the PM doesn't interface with team members daily, projects where the schedule is critical to the customer and will thus drive cost, and projects where the team is spread over multiple locations. Collecting reports of schedule progress weekly will send a message to the entire team that schedule is important. Collecting schedule status in a weekly face-to-face meeting with teammates present is the best way to get real understanding of performance issues.

The most common method of reporting schedule status is the planner/scheduler putting collected status into a tool like Microsoft Project and seeing what it does to the critical path and downstream milestones (including project completion). This is usually done monthly, even when status is collected weekly. Corrective actions are assigned when slips impact critical milestones. Another approach is to do milestone counting. Each task start and stop date is given a point value (e.g. one point for a start and two points for a finish). Points are then recorded for each schedule owner and compared against planned points in the baseline on a weekly or monthly basis. While this doesn't show the impact on the critical path, milestone counting focuses the team on meeting all commitments.

Some PMs feel that if the schedule is monitored and the planned dates are met, cost will take care of itself. And, for labor-intensive contracts, a strong correlation between schedule and cost does often exist. However, to avoid cost overruns, both schedule and cost status should always be monitored closely and forecasts of schedule and cost performance compared and used as sanity checks against each other.

Here is a list of common schedule control metrics:

- Schedule variance
- Milestones late (aged)
- Average duration/planned duration for completed tasks
- Actual hours vs. plan
- Schedule performance index
- Number of class hours training conducted
- Milestones completed on schedule
- Software problem reports opened
- Cumulative # of open software problem reports
- Number of design changes
- Software problem reports closed
- Number of COTS product version upgrades
- Number of requirements changes
- Number of temporary fixes for COTS products in use

If you find yourself behind schedule, you may need to impose seemingly drastic measures to control the schedule performance. Such things as extended workweeks, added staff, weekly or even daily schedule reviews, and planning with daily "inch-stones" can be effective in improving schedule performance. Of course, any schedule impacts should be immediately communicated to management.

One key to schedule success is to explain the project activity network to the entire team and help them understand the downstream effect of being late on a small milestone. If this is done effectively, then all Project Team members become schedule focused and meet their individual schedules or ask for help. Try to create an atmosphere that encourages honest project status from your team members. This is the only way to avoid unpleasant surprises.

## 4. **Change management (CM)**

One of your major functions is to ensure that the project performs all the work and produces the deliverables and ONLY the work and deliverables required and documented in the contract and specifications. Any deviation from what appear in the contractual scope, service delivery or performance documents is considered change and must be handled using the change control process. The key elements of this process are document and software release management and baseline management.

Depending on the complexity of your project, you can use change management tools like Caliber Requirements Management or Borland Star Team Enterprise Management systems unless your client or organization specifies a different configuration management system. The Quality Department, if it exists, can provide you with quality management and configuration management guidance and support.

No project of even moderate complexity will be planned or executed perfectly. There is always room for improvement. One method of building continuous improvement into more complex projects is to add the role of the Configuration Control Board or CCB. Real and potential process, product and performance problems are reported to the CCB for corrective action. The CCB then uses the configuration control process of logging the problem, assigning action and tracking progress in solving these problems.

*Change control board (CCB)*

Change is inevitable on projects. Customer requirements, funding, and schedules may change. Rates (labor, overhead, and G&A) may change. Approaches selected in the proposal phase may not work out. Key personnel may not be available when needed, or not at all. In these situations, the PgM Plan must be changed to reflect the reality of the situation, evaluated against the contract for scope change and coordinated with the customer. While change is necessary, successful projects manage change carefully. Unnecessary or unaffordable change often leads to disaster.

*Configuration and change management process*

The next figure illustrates the basic configuration and change management process that should be implemented on your project. However, the entire CM process is deserving of much more attention than

I can give it in this Guide. I'd recommend PMs review the PMBOK® configuration management process guidance at pmi.org.

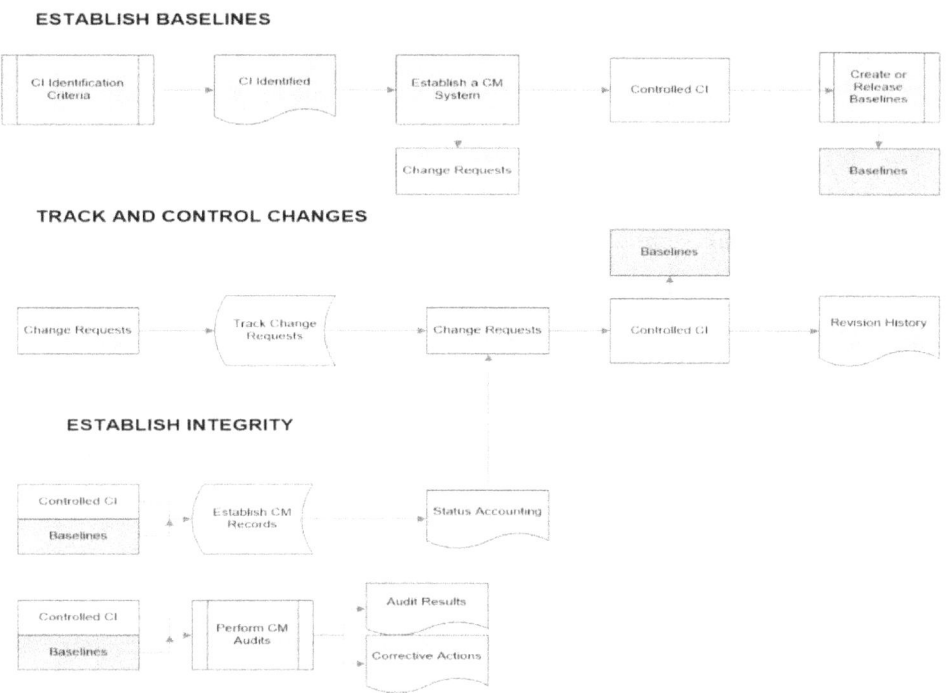

## 5. Quality management

Quality management (QM) and control (often referred to together as quality assurance or QA) involves monitoring the project and its progress to determine if the quality standards defined during Planning are being implemented and whether the results meet the quality standards. The entire organization has responsibilities related to quality, but the primary responsibility for ensuring that the project follows its defined quality procedures ultimately belongs with the PM.

Poor quality results in increased costs, low morale and low customer satisfaction. High quality results in lower costs, engaged and productive Project Teams, high customer satisfaction and lower risk.

In my experience, QM is too often neglected on IT services contracts. So, I'd recommend every PM become proficient with basic quality methods and apply them especially to the review and approval of contract deliverables. This is the reason I've provided additional detail in this subsection.

For many decades, my wife has been my unpaid Quality Control Supervisor. We keep "halving the distance to the wall," seeking that perfect document. I shudder to think what my articles and books would have looked like without her! I spared her having to edit this Guide. So, I take full responsibility and apologize for any mistakes you've discovered.

*Quality planning*

You should specify the use of quality control processes such as quality assurance of conformance to work processes, verification and validation, joint reviews, audits and process assessment. Also, clarify who will be responsible for reviewing compliance with each of these quality requirements.

For more complex system and software development projects, you should try to implement a Capability Maturity Model Integration Level 3 or higher. This is characterized by having a set of defined and documented standard processes established and subject to some degree of improvement over time. Basically, you should define the processes to be used to control, audit and measure the quality of the work and the resulting work products.

Your Quality Director can help you identify the appropriate quality elements to implement for the scale and complexity of your project.

*Implementation*

The Quality Control Manager (QCM), or you if there isn't one, should implement the Quality Plan using a systematic, step-by-step approach to ensure that the quality of services performed meets and/or exceeds contract requirements regarding timeliness, accuracy, appearance, completeness, consistency, and conformity to appropriate standards and/or specifications. The following subparagraphs explain the processes and procedures that will ensure that the performance objectives are met and/or exceeded.

*Inspection Methods to be used*

Inspection methods that can be used, but are not limited to, include:

- Personal evaluations
- Standard Operating Procedures (SOP's) specifically tailored to the location and the type of services or systems you're providing

- Checklists developed for each task
- Review of quality acceptance reports
- Review of customer complaints
- Scheduled and unscheduled quality control inspections of each functional and task area
- On-site inspection by Corporate managers, scheduled and unannounced

The primary method of identifying deficiencies in the quality of services to be performed before the level of performance becomes unacceptable should rely on a multi-tier/multi-level inspection process. This method incorporates the Total Quality Management or TQM concept known as Total Employee Involvement (TEI). TEI is simple in both theory and practice. Applied to inspections, TEI has two separate and distinct elements. The first element is actively involving all project employees in inspecting service input and output. The second element involves empowering and continually motivating employees to perform inspections of in-progress and completed work.

TEI ensures the completed work is of high quality. The roles and responsibilities for TEI are defined in this table:

| Individual | Inspection Responsibility |
|---|---|
| PM/On site Lead | Conducts project-wide, formal, documented (scheduled and unscheduled) inspections of work quality and timeliness. Reviews customer feedback/complaints. Provides constant and continuous reinforcement of the roles and responsibilities of each and every employee in the implementation of the quality control program. |
| Quality Control Manager/Deputy PM | Conducts project-wide formal, documented scheduled inspections of work quality and timeliness. Provides constant and continuous reinforcement of the roles and responsibilities of each and every employee in the implementation of the quality control program. Provides continual hands on training and education of quality control issues focused on increasing employee awareness and participation. |
| Examiner Lead and Alternate | Conducts semi-formal inspections of work quality and timeliness of their assigned functional group and task groups. Emphasis is on service output (e.g., results). Performed mainly by inspecting in-progress and completed work. |
| Remaining Employees | Conducts informal inspections of work quality and timeliness of their assigned functional responsibilities. Emphasis is on service process (e.g., input). Performed mainly by inspecting in-progress work. |

Inspections should be performed by a QCM on a weekly basis. Employees should review their work daily to assure compliance with contract requirements. The QCM or an equivalent function should perform quarterly quality reviews in addition to reviewing and QCM may also conduct unannounced quality reviews throughout the contract period of performance.

*Deficiency Correction*

Once a defect is identified, the QCM should be notified immediately. Any and all issues should be addressed immediately, thoroughly reviewed and corrected. This corrective process focuses on two steps: Root Cause Analysis and Process Control/Continuous Process Improvement. Both proactive steps eliminate reliance/dependence on Government customer identification and direction for correction of deficiencies prior to resolving the defect. It's important to note that to close the loop of this cycle, both Government and customer input/feedback should be solicited before implementing the Process Control/Performance Improvement measure(s).

*Root Cause Analysis*

After a defect has been identified through inspections, the QCM or a member of the Project Team should perform a "Root Cause Analysis." This analysis simply means that the defect is analyzed to identify the underlying procedural or systemic cause of the defect. To fully ensure identification, the QCM should employ a routine, systematic approach to problem elimination. In other words, the team should eliminate non-contributing causes/factors and methodically narrow it down so that the contributing cause/factor can be definitively identified.

*Process Control/Continuous Process Improvement*

Once the root cause of the problem has been identified, focus should then be immediately shifted to developing process control/performance improvement measure(s) that concentrate on preventing reoccurrence and thus continually improving services. The benefit of these measures is that they optimize the process and procedure by eliminating "weak link(s)" in the process or procedure. Your approach to implementing preventive and corrective actions so that they are suitable to the deficiency relies on developing a tailored/customized/case-by-case response to each problem. Examples include:

- Revised Standard Operating Procedures (SOP's)/Work Instructions
- Additional Training of Project Personnel

Once the final process control/performance improvement measure has been selected and implemented, review the area that was previously deficient to test whether the corrective action implemented will assure that a defect will not be repeated. This review is continued until the process has fully matured.

*Trend Analysis*

On more complex projects, routinely perform trend analysis as part of the QC Program. Trend analysis concentrates on identifying trends in the performance of contract work by reviewing previously compiled quality data (e.g., inspection results). This includes identifying both negative and positive trends/shifts impacting contract performance.

Once these trends are identified, you should have the authority to order the appropriate action. The nature of the action is tailored/customized to the trend(s) identified. Specifically, positive trends are recognized and acknowledged, whereas negative trends should be corrected at the procedural/systematic level to eliminate reoccurrence. The corrective action taken should have the goal of achieving a level of performance at or above contract requirements. This approach helps ensure the continuity and consistency of high-quality services. Once the appropriate action has been taken, the trend(s) continue to be monitored for adherence to or exceeding of contract requirements.

*Recording and Processing Customer Complaints*

While your objective will be to achieve a "Zero Defects" posture, complaints will sometimes occur. The QCM should receive and document all incoming complaints and then evaluate the complaint and either resolve it or relay it to the PM for resolution. In those cases that are a result of poor workmanship, the QCM will review the work, discuss the discrepancies with the affected employee and implement corrective action. The QCM ensures that the complaint is recorded and processed in accordance with contract requirements and the Quality Control Plan.

All Quality Control records concerning the performance of the contract should be maintained in the QCM's office. These files should contain inspection reports, records, discrepancies, corrective actions, copies of all incoming and completed customer complaints, etc. These records should be available for Government review at any time upon request.

*Customer Comment/Complaint Program*

On some projects, it will be necessary to provide a method for customers to report complaints, deficiencies and comments in general concerning project performance. This method is most often used for the Help Desk Support Services function of a contract. Valid customer complaints are one method by which deficiencies are identified by individuals other than internal organization or Project Staff. Where there is a case of poor performance, non-performance or a dissatisfied customer, you should thoroughly investigate the report, and if valid, document the report and take corrective action.

Through the use of the Customer Complaint Program, as well as other internal measures of Quality Control, such as employee input, scheduled and unscheduled inspections, you should be able to create an environment which will be able to not only correct current deficiencies, but will be able to identify and resolve any and all potential areas of concern prior to a deficiency becoming a reality.

## 6. Risk monitoring and control

The risk management process isn't a *one-time-and-then-forget-it* process. The risk dynamics of a project change throughout the project and risks enter and leave a project as time passes. You must continue to identify, assess and mitigate risk in order to stay on top of changes.

As the PM, you should coordinate or lead efforts to quantify, assess and develop potential risk mitigation actions for high impact/high probability project risks. This will often require the performance of a trade-off analysis to identify the most cost and schedule effective risk mitigation approach. A matrix similar to the one shown below can be used to analyze each mitigation alternative. I guess you could consider this an objectively subjective assessment of the alternatives! It takes a considerable amount of the distracting emotions out of the equation by diverting everyone's attention to the decision process.

| Criteria | Alternative 1 (Rate 1 to 10) | Alternative 2 (Rate 1 to 10) | Alternative 3 (Rate 1 to 10) |
|---|---|---|---|
| Cost impact | | | |
| Schedule impact Low/Med/High | | | |
| Risk of failure Low/Med/High | | | |
| Staff requirements Low/Med/High | | | |
| Total score | | | |

With the input from the client and industry partners, you should recommend courses of action and implement those courses of action within the constraints of time and costs as specified by the project budget. Should any risk mitigation action require significant cost or time to implement, then a written request must be developed and submitted to your manager. Any contract scope changes must be communicated by your Contracts Administrator to the client Contracting Officer.

Here are just a few examples of mitigation actions I have observed and the risks they mitigate:

- Budget management reserve   - mitigates cost risk
- Schedule slack/lag          - mitigates schedule risk
- Parallel development        - mitigates technical risk
- Propose an incentive fee    - mitigates cost risk
- Interim Progress Reviews    - mitigates cost, schedule, and technical risks

You should review the risks documented in a risk register (or log), using the following table, with senior management and, where appropriate, with the client during progress reviews and reports:

| Risk scenario | Probability of occurrence (L, M, H) | Adverse consequence (L, M, H) | Mitigation steps | Current risk status |
|---|---|---|---|---|
|  |  |  |  |  |
|  |  |  |  |  |
|  |  |  |  |  |
|  |  |  |  |  |

*L = Low   M = Medium   H = High*

Some of the common risk status descriptions are:

- Risk documented, but analysis isn't performed
- Risk analysis done, but response planning not performed
- Risk response planning complete
- Risk trigger has occurred and threat has been realized
- Realized risk has been contained
- Identified risk no longer requires active monitoring

The project duration and complexity will usually prescribe how often risks should be re-evaluated.

*Periodic Assessments* are conducted at predetermined intervals, normally during milestone reviews. This may be appropriate for projects with limited resources. However, with this approach, low risks could develop into higher project risks if not identified early enough in the project.

*Continuous Assessments* are a more proactive approach, allowing project risks to be identified early and mitigation strategies to be developed before risks impact performance, cost, and/or schedule.

*Independent Risk Assessments* are accomplished by an expert (either from a different organizational unit or an industry partner). This is especially useful when an unbiased review of project risks is needed coupled with expert recommendations.

## 7. Issue management

Issues are non-mitigated risks that have been realized! Project issues may arise from a number of sources. These include, but are not restricted to:

- Project stakeholder issues (internal or customer)
- Funding issues
- Staffing issues
- Technical issues
- Management issues

Issue analysis involves determining why a problem has occurred, as well as the corrective action required to rectify the problem. Corrective action is required when, the issue, if left unresolved, will prevent you from meeting contract requirements or the overall project objectives.

*Monitoring and Correcting Project Issues*

The PM is responsible for the ongoing tracking and monitoring of project issues. You should:

- Keep track of identified issues.
- Identify new issues.
- Evaluate and communicate any changes in issues.
- Ensure corrective actions are implemented by the due date.
- Review issue status.
- Evaluate the effectiveness of corrective actions.

The corrective action required in response to an identified issue will vary, depending on the nature of the issue. The determination of the required corrective action may require a discussion of alternative options to resolve the problem. Depending on the size of the issue, the identification of the corrective action may require more formal decision making in order to select the most appropriate solution to rectify the problem.

All issues should be tracked in an Issue Log and must be assigned an owner (the person responsible for implementing the corrective action) and a due date by which the corrective action should have occurred. Corrective actions must also be reviewed to determine if the action taken was effective in resolving the problem. This review must be conducted before the issue is closed. Where the corrective action taken was not effective in resolving the issue, appropriate action must be taken to ensure that the issue is effectively resolved.

## 8. Subcontract and procurement management

You might also be using subcontractors to provide additional support services. In most organizations a Subcontracts or Contracts Administrator will be in charge of procuring these services for you.

Depending on the type of project, you may have numerous material requirements that you and your Project Team will need to have ordered by your organization. A Procurement Specialist can walk you through this procurement process so that you don't deviate from accepted government practices.

Your primary responsibilities as a PM are to:

- Submit a purchase request (PR) for approval in the Financial System.
- Provide a subcontractor SOW (services) or bill of materials (products).
- Provide the buyer with three or more potential resources.

Please note that all procurements must be made by an authorized Procurement Professional.

It's important determine as soon as possible if any of these items fall on the project critical path. Also, account for the PR-to-Purchase Order approval timeframe.

Managing customer furnished products, information, and equipment (often referred to as GFE/GFI on Government contracts) is one of the PMs many responsibilities. Clear responsibility for accepting, maintaining, and accounting for all furnished products, information, and equipment should be assigned. However, the PM is accountable and must be cognizant of the status of all such items. In addition, the PM must ensure that customer furnished products are provided on the agreed-to schedules and in the agreed-to configurations, that documentation and software are complete and correct, and that equipment operates properly and is accurately tagged and entered into the property management system.

If furnished products are not provided on schedule, do not satisfy agreed-to configurations, and/or do not operate properly, the PM, through the Contracts Manager, must inform the customer of the problems and seek speedy resolution. The PM should also initiate planning within the project to identify alternatives to mitigate deficiencies in the furnished items.

## 9.  Human resources management

One of your most important responsibilities is to assign work to the Project Team and ensure that the work is completed according to the project schedule. A good PM establishes and maintains a Project Schedule that minimizes team member down time. Along with your Team Leaders, you must continuously communicate to each member of the team what is required and by when, and then manage the performance of each team member in meeting the requirements. Some of your human resource responsibilities are:

- Ongoing recruiting activities
- Team building activities
- Expectation setting
- Regular communications (including listening!)
- Performance reviews
- Performance evaluation and feedback
- Recognition and rewards
- Training & career development plans
- Personnel changes

For each of your employees, you should set goals and expectations, assess performance, identify development opportunities, and reward good performance and the attainment of goals. A simplified performance management timeline is illustrated in the next figure:

| January through April | | | June | Ongoing | December |
|---|---|---|---|---|---|
| Prior year performance review | Corp & group goals set updated | Ensure employee & manager goal alignment ****** Merit increase cycle | Employees & managers meet to discuss work performance against goals | Employees & managers revisit goals | Managers & employees received notifications to write self-assessments |

Some of the metrics that you can use to evaluate overall staff performance on your project are employee turnover, performance evaluation and training hours per employee.

Defining the employee's training and career development plans is also a key responsibility of PMs. It's a major contributor to staff retention. Sometimes there's the need to make personnel changes both for positive reasons such as a promotion or transfer to a new contract or as a result of performance issues. It's absolutely crucial to follow the established HR process for dealing with problem employees. Every PM should be familiar with these HR references:

- Recruiting system
- Wage & Salary Class Guide
- Adverse action review process
- Grievance review process
- ADA basics
- Preventing workplace harassment
- Preventing workplace discrimination

As a manager of people, you'll also need to familiarize yourself with your organization's adverse action and grievance review process. As a rule, you should discuss all personnel issues with a Human Resources representative prior to taking any disciplinary action either verbal or written. Also, note that all formal written counseling sessions and disciplinary actions require HR review. And, based on my own experience, when it comes to employee issues...document, document, document!

Finally, keep in mind that keeping a project team member who isn't able to improve their performance is usually not a good idea for them or the morale and productivity of the rest of your team. I believe this is the most difficult challenge every manager I know has faced. But, putting off this kind of personnel decision rarely gets easier over time and can have major negative impacts on your project and customer relationship.

## 10. <u>Communications management</u>

Communications management is the process by which important project data is identified, created, reviewed and communicated within a project. The most common method of communicating the status of the project is via a project status report. Some other communications vehicles you can select from according to your needs to ensure effective communications with all project stakeholders are:

- Status meetings
- Technical interchange meetings
- Memos
- Newsletters
- Executive correspondence
- Meeting minutes
- Executive meetings
- Steering committee meetings
- Presentations to special interest groups
- Customer focus group meetings

Communication within a project isn't only about speaking face-to-face with somebody or sending out a quick e-mail about a meeting; communication also encompasses the distribution of knowledge and information. One extremely simple way to damage employee morale and create work stoppages across divisions is for management to fail to communicate critical information and, instead, to offer only the bare minimum of information to its employees. Withholding critical information causes employees to make up their own information and begin rumor mills.

By dispersing relevant information in a timely manner to employees, you can eliminate the rumors, the loss in productivity and the needless worrying. It's critical if any project is to be successful that communication occurs effectively at all levels.

Meetings can be an efficient means of communication but they can waste time if not conducted efficiently. All meetings should have an agenda provided to the attendees before the meeting. One attendee should be nominated to maintain a record of topics discussed during the meeting, actions assigned, and any conclusions reached.

Conduct regular meetings to communicate information to the team, establish priorities, make decisions, and build morale. Periodic meetings are commonly held to resolve issues, review work products, determine work status, and consider changes to scope. The most common topic for periodic project meetings is the status of the work based on the project metrics.

A PMs job is to identify problems and fix them before they become insurmountable. Interpersonal conflicts, if left unchecked, will often negatively impact project performance. This includes your own conflicts with client or project personnel. Learn to deal with problems head-on. If you can't effectively resolve a communications conflict, then go up a level in your organization to get support.

It's my experience that when there is a significant issue on a project, there's at least one person in the organization who knows the problem exists but who is too afraid to speak up. The key question is whether you encourage honest reporting (as opposed to shooting the messenger), dig deep enough to uncover any issues and are willing to share problems with senior management. Things like this don't get better over time. So, don't bury your head in the sand!

The next table describes several useful communications vehicles and their application:

| Vehicle | Audience | Use |
|---------|----------|-----|
| Project All Hands Meetings | Entire Project Team | General information dissemination (e.g., project funding, organizational announcements, change of management, recognition of outstanding achievement). |
| Project Newsletter | Entire Project Team | Team building |
| Weekly Staff Meetings | Project Management/Leaders | Status updates, PgM Plan changes, workarounds, risk assessment, contingency plans, etc. |
| Technical Interchange Meetings | Relevant Project & Customer Staff | Specific topics such as interface agreements, design standards, system performance, etc. |
| Design Reviews | Relevant Project & Customer Staff | Establish and communicate a design baseline |
| Test Readiness Reviews | Relevant Project & Customer Staff | Assess readiness for upcoming test event. Set expectations regarding conduct and results. |
| Customer Reviews | Relevant Project & Customer Staff | Status updates, setting expectations, agreeing on PgM Plan changes. |
| Social Events | Entire Project Team | Team building |
| Email | Anyone | Informal communications with written record. |
| Teleconference | Anyone | Informal communications without record. |
| Contracts Letter | Customer or Subcontractor | Formal, official communication. |

*Project status report*

A monthly internal project status report that feeds into organizational project reviews should include the following information:

- **Status Summary** – indicating any significant impacts to the project
- **Major Accomplishments** – a list of the most important completed tasks, or a description of work done toward their completion.
- **Project Milestone Report** – a high-level glance at the major project deliverables, with their intended and actual start and end dates.
- **Issues analysis and Issue Response** – a running list of open and closed issues.
- **Change Request Analysis** – a running diary of actions taken toward acceptance of change control.

- **Risk Analysis Report** – any Risks that may be turning into a project issue and report on any situation that occurred that resulted in the Project Team being unable to perform work.
- **Financial Commentary**
- **PM's Comments** and significant planned accomplishments for the following weeks.

## 11. Customer relationship management and contract growth

Customer relationship management (CRM) is the process organizations use to optimize client communication and ensure client satisfaction. The following subsections describe ten key CRM activities:

A. Know your stakeholders
B. Understand your contract
C. Be careful what your promise
D. Manage expectations
E. Don't let your scope creep
F. Don't take things for granted
G. Don't surprise your client
H. Be proactive / offer solutions
I. CRM guidelines
J. Grow your contract

*A. Know your stakeholders*

To avoid repetition, please refer to the Section II Project Planning for additional information about project stakeholders.

*B. Understand your contract*

You should know your contract backwards and forwards. This is the only way to ensure that you deliver everything your organization is contractually obligated to provide while you avoid doing work that is outside the scope of the contract. Some of the key elements of your contract that relate to CRM are:

- "Shall" statements (what your organization must do)
- Specific deliverables and special requirements
- Tasks that are required to reach client's stated goals
- Ancillary tasks (acquiring special equipment; hiring specialists; training)
- Reporting and communicating tasks to the customer

*C. Be careful what you promise*

A majority of government contracts have a defined scope of work. Some are vaguer than others. Your responsibility as a PM isn't to deliver everything your clients want regardless of cost and schedule impact. That's why it's important to make sure you and your client have a clear understanding and agreement regarding what is in and out scope to the contract. This will be one of your biggest challenges as a PM.

This means GET IT IN WRITING!

It will be tempting to agree with your client counterpart when asked to add an additional feature, do a little more work for free, skip a documentation standard, etc. If it's important enough for you to do, it's important enough to get in writing. Myself and many other PMs learned this the hard way. So, beware of any verbal direction that impacts your contract.

*D. Manage expectations*

Some ways to manage your client's expectations are to confirm every agreement in writing; make sure to only agree to fulfill requirements you can meet within the schedule and budget; give them ample warning of any anticipated schedule slips, technical deficiencies or staff turn-over. In my experience, these steps will greatly minimize customer misunderstandings.

E. Don't let your scope creep

While you should do your best to provide outstanding service and solutions, you shouldn't strive for perfection. Rather, you should do your very best to fulfill the contract requirements in accordance with established processes and procedures.

Scope creep is a common occurrence on IT projects and is perhaps the number one cause of cost overruns, schedule delays and poor customer relations. It's important that you bring any potential contract scope expansion to the attention of your client and, coordinating through your Contracts Administrator, process the necessary change proposal to accommodate additional contract requirements before initiating new work.

*F. Don't take things for granted*

PMs can't take things for granted. Too many things can go wrong without constant attention and oversight. Your client is assuming you'll stay on top of the status of your project. This means regular status meetings and reports as well as the maintenance of the risk register described earlier in subsection 6. Don't assume that everything is on track just because no one has come to you with problems.

You should also coordinate with the IT Department regarding any changes in your project IT support needs.

*G. Don't surprise your client*

A key component of CRM is to identify and communicate potential project risks to your client before they become more serious issues. In other words, don't wait until the day before a deliverable is due to tell your client you're going to miss the deadline.

*H. Be proactive and offer solutions*

No one -- your supervisor or your client -- wants to just be handed a problem. Part of the key to being a successful PM is to research and present a solution or a set of alternatives. In more complex cases, the appropriate problem reporting communication includes giving your client a schedule of when you'll complete the data gathering and analysis necessary to identify the optimum solution.

*I.  Learn how to deal with common CRM challenges*

Here are lessons I've learned regarding how to deal with some of the more common CRM challenges:

- You and your supervisor should establish a rapport with the key client stakeholders. This means visiting your client on a regular basis and your supervisor visiting the next level client management on a monthly or bi-monthly basis to seek feedback on project performance and to help the client identify new opportunities for your organization's support.
- Make sure to conduct thorough kick-off meetings both internal and then with your customer.
- Establish and document project objectives in your PgM Plan.
- Make sure your client buys into your technical approach.
- Don't let your PgM Plan gather dust. Instead refer to it on a regular basis.
- Measure your project progress against the plan and modify it where necessary to reflect project baseline changes.
- Establish a collaborative win-win partnership with your client.
- Help your client understand that your organization is a services and solutions delivery company as opposed to a personal services company.
- You deliver high quality services and products and, for instance, want to avoid your client managing your staff's timecards and sending your staff on personal errands.
- Help your client focus on contract deliverables and milestones instead of the details.
- Understand your client's success criteria and establish corresponding project success measures (like a balanced scorecard).
- You and your client should know what constitutes success. If you don't, you'll never be done! This is especially important on fixed price tasks.
- Try to set expectations that you can meet or exceed.
- Read your contract...especially the SOW...and refer back to it before agreeing to additional tasking.
- You're responsible for controlling the scope of work and changes to it! Ignore the contract at your peril.
- Regularly review with your stakeholders the project technical, cost and schedule status compared to client expectations.
- Actively seek client feedback regarding what is going well and not well. Also, listen to your teammates regarding project status and concerns. You can't fix it if you don't know about it.
- Communicate workarounds or assumptions to your client in writing.

- You need to be willing to say no to your client:
  - When something the client wants is out of scope
  - When a request is unreasonable or unethical
  - When something isn't in their best interest, you can say, "Our experience has shown us that…"
- When you must say no, try to offer a more attractive alternative to solve client's problem. When all else fails…talk to your manager!
- The truth is easier to remember a month later than the lie you told yesterday!
- Don't always give the client everything they ask for or want. This will require you to build a strong rapport and mutual trust.
- Document disagreements, especially regarding financial and accounting issues. Alert upper management promptly to any unresolved issues or client dissatisfaction.
- The quality of your relationship with your client can be an excellent barometer of the status of your project. Make sure not to avoid your client. Even bad news needs to be communicated in a timely manner to avoid doing serious damage to your credibility and relationship.
- To get your project out of trouble:
  - Meet with your clients to understand their concerns – this means really listen to what they have to say.
  - Bring a director or VP to show commitment or when you have to deliver bad news (like you can't spend more of your organization's $ on fixing a problem that isn't your organization's fault.
  - When necessary, work with your client to re-define project closure. Make sure to include your Contracts Administrator in this process and then make sure any baseline changes are documented.
- All employees are expected to behave in an ethical manner. There is absolutely no exception to this. If in doubt, seek guidance. If you're uncertain about your client's business practices, talk to your Contracts Administrator or upper management.

*J. Grow your project*

Once you have successfully kicked off your project, you should be on the lookout for additional tasking both with your existing client and other key client managers.

The other major business development activity every PM needs to be aware of is the need to perform early re-compete capture activities. Of course, maintaining customer satisfaction is a major element of winning a re-compete. But, waiting until the last minute to prepare and execute a capture plan is a recipe for disaster.

At least several months before your contract is to be re-competed, you should define and implement the winning capture strategy. Be a champion for necessary capture and proposal resources. Understand the client's needs and biases. For instance, is someone in the client's organization or a small business contractor pushing to have your work set-aside for small business competition? And, if so, can you or your organization's representative make a convincing argument why this acquisition change would jeopardize contract performance? As the PM, it's your responsibility to keep the organization focused on the capture of your re-compete.

Every re-compete should have some level of capture plan. This plan should include:

- Acquisition strategy
- Competitive analysis
- Project improvement actions
- Positioning
- Customer requirements changes
- Design to cost
- Teaming/subcontract changes
- Pricing strategy including determining the most likely Price-to-Win range

In general, the re-compete process involves understanding the client's acquisition strategy, the potential competition including who is teaming together and their strengths and weaknesses. Also, as the incumbent, how you can correct any of your weaknesses prior to the acquisition process start. Depending on the size and complexity of a project, the capture manager for your re-compete might be you as the current PM, an assigned capture manager or an organization executive.

Re-competes should follow your organization's BD and bid decision process and utilize the sales lead tracking system.

The first consideration is to determine whether your organization wants to continue to perform this work. Factors to consider include:

- Has it been profitable?
- Is the work doable? Is the customer too difficult to work with?
- Does the client want you to win again (are they satisfied with your support)?
- Does it take too much management attention to oversee the project for the realized revenue and profit?
- If a very small task, does it have sufficient expansion potential to justify bidding again or would you better off subcontracting or handing off the work to a small business partner?

Assuming the answer is yes, some key capture positioning questions you should consider are:

- Is it time to replace any key staff (does the client like them)?
- Do you still have the right subcontractors?
- Are there any weaknesses that need to be corrected?
- Can you make these changes prior to the proposal period?
- What are the key themes and discriminators?

Too often, PMs assume because the client is satisfied with your performance, that they'll give your organization high proposal scores. However, this assumes your key clients are on the proposal evaluation team. This isn't always the case.

Make sure you describe the features and benefits of your technical approach even if the client knows this already. Give them the information they need to be able to score you highly in the proposal evaluation.

Describe your management approach including staffing, your organizational structure and task management process. Include a management risk assessment and mitigation discussion to demonstrate your understanding of the job. This is a proposal technique that can really discriminate an incumbent from the competition.

Regarding pricing, consider whether you need to be more aggressive in the pricing of the offer. There will most likely be one or more other companies bidding an aggressively low price hoping to unseat you as the incumbent. Sometimes this means you'll need to replace some personnel and hire more junior staff (except for key personnel the client loves). Don't just assume the client will pay a premium to keep you around. Also, do you know what this $ delta is? In other words, avoid drinking your own bath water! In today's lowest-priced-technically-acceptable bidding environment, complacent contractors rarely win follow-on business.

## 12. Security

As the PM, you're responsible for all aspects of your program to include Security. Your government counterpart will have a Security Department supporting them that will interface directly with your Security staff and perform inspections. Introduce your Security Team to your customer's Security team and let them work together.

When issues, such as a Security incident arise, you'll become part of the problem resolution process since it's your personnel and program that are impacted. Some of your Security responsibilities are:

- Ensure all personnel comply with Security requirements of the contract to include reporting changes in personal status, foreign travel, foreign contacts, etc.
- Work with Security to ensure all contact requirements are addressed and completed.
- Report any Security incidents immediately to Security even if your customer has taken the lead on the investigation.
- Coordinate any changes to the contract or procedures.

Normally, HR will notify Security of changes such as reassignments or terminations. When you add somebody to your program, you may have to get your customer's approval before Security can transfer Security clearances. This action has can take up to 90 days or longer. Polygraphs are required by some of your customers and many times the polygraph is one of the most intimidating phases of Security to your personnel. Your Security Team probably has a polygraph pre-briefing video and can discuss polygraph issues with you or your team members before they take one.

Communications is a two-way street. Security has an obligation to inform you of any changes the customers makes that might impact the Security posture of your program.

# Section 4: Project close-out

The Project Close-out Phase is the last phase in the project lifecycle. Closeout begins when the government accepts the project deliverables and concludes that the project has met the goals and requirements established in your contract. The major focus of project closeout is:

1. Contract close-out
2. Subcontract close-out
3. Finance close-out
4. Deliverable close-out
5. Project records close-out
6. Staff close-out
7. Lessons learned documentation
8. Security close-out

Appendix E contains a sample project closeout checklist.

The following table summarizes the key information contained in this section:

| Tasks | Deliverables (Outputs) | References/ Tools | Interfaces | Key Points |
|---|---|---|---|---|
| 1. Contract close-out | -Final contracts documents | -FAR 42-4.804 -DD Forms 1594 (Contract Completion Statement) & 1507 or 1597 (Checklists) -Procurement Defense Desktop (PD2) | -Contracts Administrator | -Review contract milestones, deliverables & close-out contract requirements -Return any Government Furnished Property (GFE) -Archive project documentation |
| 2. Subcontract close-out | -Subcontractor performance report -Subcontractor final invoice -Subcontract Close-out Checklist & Certification Form | FAR Part 44 | -Subcontracts Administrator | -Determine that all requirements have been met -Obtain all your organization's property -Provide Subcontracts Administrator with performance report (Word document) -Archive subcontractor documentation |
| 3. Finance close-out | -Client project completion approval -Action items completed | -Financial System | -Finance | -Obtain cost information from finance & accounting -Finalize expense reports -Conduct inventory of capital equipment -Return GFE -Close WBS #'s and Activity IDs |
| 4. Deliverable close-out | -Obtain project completion agreement from client, corporate management & team | | -Client -Corporate management -Contracts -Subcontracts | -Validate metrics -Assess deliverables -Obtain completion approval -Complete open action items & verify item closure |
| 5. Project records close-out | -Project Notebook -Concept document -Project Charter -PgM Plan -Correspondence -Meeting notes -Status reports -Contracts & subcontracts files -Tech documents, files, programs, tools | | -Contracts -Subcontracts -Security | -Coordinate with Security for return/storage/destruction of classified data & artifacts |
| 6. Staff close-out | -Staff performance evaluations | -Your organization's Human Resources Policies | -Human Resources | -Get full time staff back into available resource pool as quickly as possible -Communicate employee performance to functional manager -Make recommendations regarding performance recognition -Make sure each employee's project hours have been correctly accounted for |
| 7. Lessons learned documentation | -Lessons learned recommendations report | | -Contracts | -Conduct lessons learned session -Document major project issue recommendations |
| 8. Security close-out | -Close-out plan with customer | -Customer Policy and Implemen-tation guidance -DD-254 | -Security | -Classified data destroyed, archived, or transferred. -Facilities, IT systems decertified -Personnel debriefed or clearances terminated |

## 1.  Contract close-out

As a PM, there will likely be pressure for you to move on to the next project. However, contract close-out is very important and shouldn't be neglected. It's a simple process, but close attention should be paid so that no room is left for liability. In order to close a contract it's important to collect all of the pertinent documentation for review and archive. This will include all of the original contracts and supporting documentation such as schedules, contract changes and performance reports. This documentation needs to be reviewed thoroughly to ensure there are no unrealized contract issues that could result in contractual liability. Your Contracts Administrator can explain the required contracts archive procedures.

A thorough review of the procurement and contracting documents should include contract milestones and services provided or deliverables and documentation delivered. Standard verbiage for acceptance and closure is usually found in the original contract itself.

The Contracts Administrator will execute final contracts documents and return them to the client.

The Federal Acquisition Regulations (FAR) specify contract closeout requirements. In addition, the Department of Defense has specific project closeout requirements that may include a *DD Form 1594* (Contract Completion Statement) and a *DD Form 1507 or 1597* (Contract Close-out Checklist). Electronic closeout processes are also now being accomplished in *Procurement Defense Desktop* (PD2). These closeout requirements should have been identified in your contract.

## 2.  Subcontract close-out

Initial responsibility for close-out of a subcontract resides with the PM. The FAR specifies overall subcontract closeout requirements. You should review the subcontract to determine that all deliverables have been submitted. The primary sources for this information are the SOW, as stipulated in the subcontract, and the submitted reports.

All subcontracts should require the subcontractor to label its final invoices as "FINAL" and to submit the final invoice within a reasonable period of the project end date. This allows for timely closeout of the subcontract and is especially important as many subcontracts are scheduled to end on the same day as the prime award. However, the final invoice shouldn't be paid until the subcontractor has met all subcontract requirements.

Make sure that all your organization's property has been returned by the subcontractor and the proper forms filled out.

When you have determined that the subcontractor has met all subcontract requirements (i.e., all reports, invoices and property have been received), submit a Subcontract Close-Out Checklist and Certification Form to the Subcontract Administrator. This provides assurance that all subcontracted work has been satisfactorily completed and provides a checklist for the Administrator to help ensure that all closeout actions have been completed.

Your last responsibility is to submit a subcontractor performance report to the Subcontracts Administrator who will archive it for future reference and government past performance queries.

## 3. Finance close-out

Financial closure is the process of completing and terminating the financial and budgetary aspects of the project. Financial closure includes both (external) contract closure and (internal) project account closure.

You'll need to obtain cost information from the Finance and Accounting Departments. All expenditures must be accounted for and reconciled with the project account. When financial closure is completed, all expenditures made during the project will have been paid as agreed to in purchase orders, contract or inter-agency agreements.

You should conduct an inventory of any capital equipment and return it to the IT or Facilities Department.

All activity IDs in the Financial System for the contract or specific task order should be closed.

## 4. Deliverable close-out

The basic steps to closing out contract deliverables are:

- Validate final actual cost, schedule, risk, engineering and quality metrics.
- Determine whether all project deliverables are complete, accurate and have been delivered.
- Obtain agreement from customer, corporate management and the Project Team that the project is complete.
- Assign staff to complete any remaining action items.
- Verify that action items are completed correctly.

## 5. Project records close-out

Verify that all project records are complete and have been provided to your Contracts Administrator for archiving. Make sure to coordinate with Security for the return/storage/destruction of classified data and artifacts in accordance with requirements specified in the contract and your organization's Security policy manual.

Some of the records that should be archived are:

- Project deliverables
- Technical data package
- Project files
- Project concept document
- Project Charter
- PgM Plan
- Project management and oversight review records
- Correspondence
- Meeting notes
- Status reports
- Contracts and subcontracts file
- Technical documents, files, tools, etc.

## 6. Staff close-out

If personnel have been committed to the project full-time, it's important for you to get these people back into the available resource pool as quickly as possible and before these individuals look outside the organization for other opportunities. This will ensure that the staff stays busy and that other projects within your organization do not fall short of resources.

In some cases, employee performance reports or other documentation must be prepared for personnel assigned to the PM. In matrix situations, the PM should communicate information about the performance of each employee to the functional manager. The PM should also make recommendations for recognition of performance as each case may warrant.

Before any employee is officially transferred, the PM or his representative must ensure that all project materials and property are turned over by the employee. You must also ensure that each employee's project hours have been accounted for and correctly charged to the project.

## 7. Security close-out

Too often the Security aspects of a project close-out are ignored or overlooked. Depending on the nature of the contract, Security is often required to support the contract until final closeout by DCAA.

As the contract comes to a close, you should work with Security to determine the final disposition of all classified media and IT systems, clearance status of all personnel and secure facilities. Customers and contracts may have different close-out requirements. The following are items you need to address so you can certify to the customer that you have satisfied all Security requirements:

- All personnel have been debriefed from their clearances or accesses.
  - Clearances and accesses are tied to a specific contract. When an employee is removed from a contract and no longer has a need to have access to classified material, the employee will be removed from access and debriefed.
  - Clearances and/or accesses may or may not be transferrable to other contracts or customers.

- All classified information is the property of the U.S. Government and must be returned to the U.S. Government upon request. Classified information isn't the property of your organization.
- All classified media (defined as documents, hard drives, other storage media, etc.) must be accounted for and either destroyed or returned to customer. Seek guidance from your Security Team as each agency has provisions to allow the retention of classified data for specific periods of time.
  - Most media will either be destroyed or returned to the customer.
  - Some media may be retained by your organization with customer approval. Some examples could be company proprietary or independent research and development (IRAD) information that is also classified, classified proposals, or information required for a follow-on contract.
- Classified computer networks supporting the contract will be decertified, sanitized and disposed of as required by the approved IT Security plan and customer guidance. In some instances the network may be reassigned to support another contract.
- Facilities used for classified activities will be closed or the Security cognizance transferred to another contract.

It's imperative you accomplish these tasks at every close-out to keep your classified holdings to a minimum and comply with contractual Security requirements.

## 8. <u>**Lessons learned documentation**</u>

I believe an especially important project close-out PM responsibility is to conduct a lessons learned session and document the results. Capture the best practices, lessons learned and the project environment for use by your organization on other similar projects.

In addition to communicating the closure of a project in writing, it's also advisable to have a mechanism for group review. Lessons learned sessions are valuable closure and release mechanisms for team members, regardless of the project's success. The lessons learned session is typically a meeting or a series of meetings that may include the following:

- Project Team
- Stakeholder representation—including external project oversight

- Executive management
- Maintenance and operation staff

For a lessons learned session to be successful, the problems encountered by the Project Team must be openly presented. It's important, however, that the problem discussions do not merely point a finger at some target other than the Project Team; responsibility and ownership for problem areas are critical to developing useful recommendations for future processes.

Problems that were encountered should be prioritized with a focus on the top five to ten problems. It's not necessary to document every small thing that happened. However, all legitimate problems and issues should be discussed. And, where appropriate, flagged to management for future action. Also, project problems should not only be collected but shared with all the PMs in the organization. This is one of the best ways to avoid making the same mistake twice!

# Appendices

## Appendix A: Project management plan template

> *The following provides information on the requirements for establishing and maintaining the PgM Plan document*

Audience
: The Project Management (PgM) Plan is intended for managing a project. It provides a ready reference for the PM, all project personnel including the Contracts Administrator, Financial Analyst (Project Control Analyst), and supervisors executing tasks on the project.

Owner
: The PM (or in some cases a Program Manager) owns the PgM Plan and is responsible for establishing and maintaining the plan.

Applicability
: The PgM Plan should be completed for all your organization's projects. The size and complexity of projects may necessitate tailoring of the PgM Plan.

Review
: The PgM Plan is reviewed by a Program Manager and/or Operational Director and, in some cases, the client, as well as by any designated project personnel or other stakeholders.

Approval
: The Operational Director approves the initial version of the PgM Plan. Updates to the PgM Plan do not require Operational Director approval unless otherwise stated.

Storage
: The PgM Plan is stored in an electronic project repository as indicated on the Project Profile page of the PgM Plan.

Update
: The frequency of update for this plan (see following table) depends upon the content that is maintained within the plan vs. the amount of content that is simply referenced. Some plans may need updates on a monthly basis while others may be updated only quarterly.

| PgM Plan Section | Recommended Update Frequency |
|---|---|
| Project Profile | Define initially then update as needed to reflect changes, review on a periodic basis |
| Project Objectives | Define initially then update as needed to reflect changes, review on a periodic basis |
| Technical Approach | Define initially then update as needed to reflect changes; review on a periodic basis |
| Project Management Approach | Define initially then update as needed to reflect changes; review on a periodic basis |
| Resource Planning | Define initially then update as needed to reflect changes; review on a periodic basis |
| Project Personnel Assignments | Define initially then update as needed to reflect changes; review on a periodic basis |
| Project Schedule and Milestones | Weekly or Monthly; frequency depends upon project duration, complexity of schedule, and number of deliverables or milestones |
| Project Monitoring (Meetings) | Define initially then update as needed to reflect changes; review on a periodic basis |
| Financial Tracking | Maintained elsewhere; financials are typically updated monthly if not more frequently |
| Procurement and Supplier Management Plan | Define initially then update as needed to reflect changes; review on a periodic basis |
| Project Metrics | Track monthly or as dictated in your organization's Metrics Plan |
| Project Completion | At interim stages if applicable, but minimally at the completion of the project |

## Key Definitions

| | |
|---|---|
| Project | The term "project" is used synonymously with the term "task". A task is funded work issued by a customer through any of a number of methods: task order, delivery order, technical instruction, or technical direction letter. |
| Work Category | Work executed on projects is categorized in one of three ways: Level of Effort, Engineering Services, or Developmental. A complete description of these categories with examples might be provided in your organization's Tailoring Guidelines. |
| PgM Plan (PgM Plan) | The PgM Plan is a PgM Plan used to manage developmental and engineering services and development projects under a particular contract. |

**Note: Remove these instruction pages once PgM Plan draft is complete**

# PgM Plan Template

<Contract Name>

<Customer>

<Project Name>

<Date>

| Document Change History | | |
|---|---|---|
| Version: 1.0 | Description of Version: Initial draft of PgM Plan | |
| | Name | Date |
| Authored/Edited By: | | |
| Reviewed By: | | |
| | | |
| | | |
| | | |
| | | |
| | | |
| | | |
| Approved By: | Operational or Engineering Director | |

<Organization Name> Confidential – Internal Use Only

*(Note: Should be marked on all pages.)*

## PROJECT PROFILE

**Instructions**: *Complete the general information about the project in this section>*

| | |
|---|---|
| Contract Name | |
| Contract Number | |
| Project Name | |
| Financial System Project ID/ Number | |
| Customer Your Organization Supported | |

| Organization Points of Contact *<Indicate key points of contact>* | |
|---|---|
| PM | |
| Project Control Analyst | |
| Contract Administrator | |
| QA Representative | |
| Security Officer | |
| Other… | |

| Customer Points of Contact *<Indicate key points of contact>* | |
|---|---|
| TPOC name | |
| TPOC phone | |
| TPOC email | |
| Project Lead | |
| SME | |
| Other… | |

| Project Repositories *Indicate repository information; delete those not applicable* | | |
|---|---|---|
| **Name** | **Location** | **Address** |
| Contract Management System | | |
| Project Online Depository | | |
| | | |
| Other … | | |

## PROJECT OBJECTIVES

**Instructions**: *Describe the work to be performed. Synopsize the project Statement Of Work (SOW).*

| Overview of Project Work: |
|---|
| |

## TECHNICAL APPROACH

**Instructions**: *Decompose the SOW into a list of requirements to accomplish the work. Establish the work categories and tasking type for the project. Referring to your organization's Tailoring Guidelines, discuss the types of plans that will be used to manage the project and the associated tasks.*

**Detailed Requirements:**

**Work Breakdown Structure**: (Note: required for all development projects, optional for all others)

**Design and Methodologies**: (Note: required for all development projects, optional for all others)

## PROJECT MANAGEMENT APPROACH

**Instructions**: *Describe the approach toward managing the project. Provide an organization chart for the project showing the chain of responsibility from the perspective of both your organization and the Customer.*

**Management Approach for the Project:**

**Notional Project Organization Chart**: Insert here.

## RESOURCE PLANNING

### Hiring Plan

| Personnel Labor Category | Number Personnel | Timeframe | Recruiting Support Needs |
|---|---|---|---|
|  |  |  |  |
|  |  |  |  |
|  |  |  |  |
|  |  |  |  |
|  |  |  |  |

### Additional Resource Requirements

| | |
|---|---|
| Equipment |  |
| Software/ Hardware Tools |  |
| Other Tools |  |
| Government Furnished Equipment |  |
| Facility (Conference rooms, SCIF, etc.) |  |
| Equipment (Projection Systems, Teleconference, etc.) |  |
| Project Repositories |  |
| Security Requirements |  |
| Training |  |

## EXPORT ISSUES

| Yes | No | |
|---|---|---|
|  |  | Will you send or take defense articles out of the United States, in any manner? |
|  |  | Will you transfer registration or control/ownership to a foreign person of any aircraft, vessel, or satellite covered by the United States Munitions List? |
|  |  | Will you disclose or transfer technical data to a foreign person whether in the United States or abroad? |
|  |  | Will you perform defense services on behalf of, or for the benefit of, a foreign person, whether in the United States or abroad? |

## PROJECT PERSONNEL ASSIGNMENTS

**Instructions**: *Identify the Project Staff and their roles and responsibilities. If they have been assigned specific tasks, indicate that informational.*

| Name | Role | Responsibilities |
|------|------|------------------|
|      |      |                  |
|      |      |                  |
|      |      |                  |
|      |      |                  |
|      |      |                  |
|      |      |                  |
|      |      |                  |

## PROJECT STAKEHOLDER ANALYSIS

**Project Stakeholders** *Instruction: Complete a Stakeholder matrix. Add the appropriate roles; include internal and external stakeholders. The matrix below is only a representative example.*

*Key: Owner, reviewer, control, approval, input, information only*

| Role | Name & title | Role | Who in your organization is the key contact |
|------|--------------|------|---------------------------------------------|
| **Your senior management** |  |  |  |
| **Your contracts** |  |  |  |
| **Your finance** |  |  |  |
| **Your security** |  |  |  |
| **Your IT** |  |  |  |
| **Client PM** |  |  |  |
| **Client technical** |  |  |  |
| **Client contracts** |  |  |  |
| **Client finance** |  |  |  |
| **Client operations** |  |  |  |

## RISK IDENTIFICATION AND MITIGATION

**Instructions:** *Indicate technical risks that affect the project. The risk cube shown is color coded according to rating guidelines (green, yellow and red). If your project uses a different standard, replace this cube with one that shows the appropriate color-coding. Include the rating criteria that your project uses to assign levels of probability and impact. These differences should be part of your approved project tailoring.*

| Risk Management Program | |
|---|---|
| Risk Management Strategy | |
| Risk Management Lead POC | |
| Risk Management Repository | |

*Select risk priority based upon likelihood and consequence.*

| Likelihood (Probability) | |
|---|---|
| 1 | Not Likely |
| 2 | Low Likelihood |
| 3 | Likely |
| 4 | Highly Likely |
| 5 | Near Certainty |

| Consequence (Impact) | |
|---|---|
| 1 | Minimal |
| 2 | Some |
| 3 | Moderate |
| 4 | Significant / High |
| 5 | Critical / Severe |

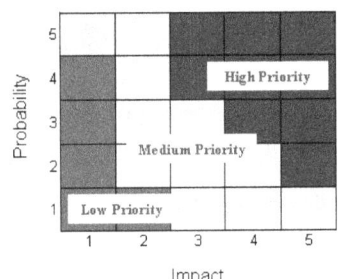

| Risk ID | Risk Title | | Date Opened | Priority | Status | Evaluation Frequency | Next Evaluation |
|---|---|---|---|---|---|---|---|
| | | | | | | | |
| | Description | | | | | | |
| | Mitigation | | | | | | |
| Date | Step | | | | | | |
| | Step 1 | | | | | | |
| | Step 2 | | | | | | |
| | Step 3 | | | | | | |
| Risk ID | Risk Title | | Date Opened | Priority | Status | Evaluation Frequency | Next Evaluation |
| | | | | | | | |
| | Description | | | | | | |
| | Mitigation | | | | | | |
| Date | Step | | | | | | |
| | Step 1 | | | | | | |
| | Step 2 | | | | | | |

# PROJECT SCHEDULE AND MILESTONES

***Instructions:*** *The project schedule should cover the current period of performance and include the following: project monitoring and tracking activities, deliverables (CDRL items) and when they are due, customer meetings, milestone reviews, senior management reviews, audits, peer reviews of deliverables, planned review and updates to this PgM Plan, and any other key dates. Change the names of events listed below to reflect those for the project.*

| Start_____ End _____ <include year> | JAN | FEB | MAR | APR | MAY | JUN | JUL | AUG | SEP | OCT | NOV | DEC |
|---|---|---|---|---|---|---|---|---|---|---|---|---|
| **Your Organization** | | | | | | | | | | | | |
| Project Review / Project Meetings | | | | | | | | | | | | |
| Peer Review <Deliverable Name> | | | | | | | | | | | | |
| Peer Review <Deliverable Name> | | | | | | | | | | | | |
| Periodic update to this PgM Plan | | | | | | | | | | | | |
| Other… | | | | | | | | | | | | |
| **Customer** | | | | | | | | | | | | |
| Customer Meetings | | | | | | | | | | | | |
| Project Milestones | | | | | | | | | | | | |
| Deliverable A due | | | | | | | | | | | | |
| Deliverable B due | | | | | | | | | | | | |
| Deliverable C due | | | | | | | | | | | | |
| Customer Final Acceptance Review | | | | | | | | | | | | |
| Other… | | | | | | | | | | | | |

| PROJECT MONITORING | | | |
|---|---|---|---|
| PLANNED MEETINGS (Technical and Other) | | | |
| Meeting Name | Purpose | Frequency | Stakeholders |
| | | | |
| | | | |
| | | | |
| | | | |

| FINANCIAL TRACKING |
|---|
| Funding and spending on this project are tracked on a monthly basis (if not more frequently) by the Project Control Analyst and the PM. This information is housed in a separate location and isn't addressed in this plan.<br><br>Location of Financial data: _____ |

| PROCUREMENT AND SUPPLIER MANAGEMENT PLAN | | | |
|---|---|---|---|
| *Instructions: If Suppliers are used on this project, complete this plan to provide the details of their involvement. Include all tasking, deliverables, and reporting responsibilities.* | | | |
| **Project Partners**   *<Check all that apply>* | | | |
| Yes | No | Question | Action Required |
| | | Will Vendors be used to make material purchases for this project? | If Yes, follow your organization's Procurement Manual |
| | | Are Subcontractors working on this project? | If Yes, complete the information below for *each* Subcontractor. Follow your organization's Procurement Manual to secure a Purchase Order.<br><br>Note: Development projects may require a Subcontractor Management Plan. |
| | | Are Consultants working on this project? | If Yes, complete the information below for *each* Consultant. Follow your organization's Procurement Manual to secure a Purchase Order.<br><br>Note: Development projects may require a Subcontractor Management Plan. |

| *Instructions: Complete the following for each Subcontractor or Consultant on the project.* | | |
|---|---|---|
| **Subcontractor / Consultant** | **Role/Contact Information** | **Area of Responsibility / Project Work Assignment** |
| Key Personnel | | |
| Key Personnel | | |
| **Subcontractor / Consultant** | **Role/Contact Information** | **Area of Responsibility / Project Work Assignment** |
| Key Personnel | | |
| Key Personnel | | |

## PROJECT METRICS

**Instructions**: *PMs should consider collecting and analyzing relevant metrics as listed below.*

| Measure Name | Purpose | Short Description |
|---|---|---|
| Schedule Variance | Monitor schedule adherence | Actual vs. Estimated days to milestone date |
| Supplier Schedule Variance | Monitor supplier schedule adherence | Actual vs. Estimated days to milestone date |
| Lifecycle Requirements Changes | Monitor requirements volatility | Number of changes by category during phases |
| Supplier Requirements Changes | Monitor tasking changes given to subs | Number of changes by category during phases |
| Project Review Activity | Determine effectiveness of peer reviews | Number of planned vs. actual peer reviews |
| Effort by Phase | Determine % effort spent in each phase | Time spent on planning, design, implementation, test, and/or maintenance |
| Peer Review Defects | Quality of deliverables | Number and type of peer review defects |
| Risk Monitoring | Effective risk management | Number and priority of risks tracked; status |
| Project Performance Evaluation | Determine customer satisfaction | Rating of performance on the project |
| Supplier Evaluation | Monitor supplier performance | Supplier performance rating in established areas |
| Training Execution | Monitor training requirements | Approved training events planned vs. executed |
| Work Product Size Estimation | Improve estimation accuracy | Basis of estimation for work product size |
| Cost Estimation | Improve cost estimation accuracy | Cost vs. Budget monitoring |
| Supplier Cost Variance | Monitor control of supplier costs | Supplier costs / expenditures vs. budgeted |
| Funding/ Expenditure Monitoring | Monitor project expenditures | Total funding vs. expenditures to date |
| Work Allocation | % your organization vs. sub work on project | % your organization's direct labor hours on the project |
| Revenue Lost (Unfilled Requisitions) | Maximize direct labor utilization | Unused funding |
| Test Defects | Product high quality products and services | Number and type of defects found during lifecycle phases |
| Acceptance Defects | Monitor defects found at acceptance | Number, type, and severity of defects found |
| Build Errors | Minimize configuration build errors | Number of errors found during product builds |
| Review adherence | Promote timely personnel reviews | Date required for review vs. actual review date |

| Location of Project Metrics |
|---|
| **Measurement and Analysis data repository**: _____ |

## PROJECT COMPLETION ACTIVITIES

*Instructions*: *Enter pertinent information throughout the project lifecycle. Provide summary information at the conclusion of the project.*

### Significant Accomplishments, Awards, and/or Recognition
*List any notable achievements, customer recognition etc. on this project.*

| Person / Event | Accomplishment, recognition, award, etc. |
|---|---|
| | |
| | |
| | |
| | |

### Process Evaluation    *For processes utilized on this project, identify suggested changes, improvements, or exemplars.*

| Process Name | Suggestion, comment |
|---|---|
| | |
| | |
| | |
| | |

### Supplier Evaluations    *List the evaluations completed for this project, if any.*

| Date | Company Name |
|---|---|
| | |
| | |
| | |
| | |

### Key Take-Aways / Lessons Learned

| Title | Synopsis | Recommendation for Future |
|---|---|---|
| | | |
| | | |
| | | |
| | | |

# Appendix B: Project start-up checklist

*Negotiate and Definitize the Contract*

- Ensure that your organization's contracts management is in concert with the technical management team (including higher management as appropriate) on all negotiation issues.
- Document and retain all negotiation resolutions.
- Ensure that the project and the customer have identical copies of the results of all negotiated items.
- Ensure that the contract, where applicable, has been updated to reflect negotiated items.
- In some situations, only with management approval, work is started before a contract is awarded. A formal funds authorization is required to expend funds.

*Conduct Evaluation and Planning*

- Thoroughly review the entire contract with any negotiated items to ensure that the approach(es) to be taken is consistent.
- Where discrepancies are found, document them, track to closure and communicate all intended changes to management for review and approval.
- Meet with the cognizant BD or sales executive to review: the status and outcomes of past discussions with the customer; your understanding of the customer's issues, needs, and environment; and any assumptions made in developing the project proposal.

*Update PgM Planning*

- Conduct a project review to establish team understanding and commitment.
- Understand the system's development environment.
- Understand specific contract requirements specifying:
- Configuration management and control
- System/software/hardware quality.
- Understand approaches proposed to achieve the project's objectives.
- Review, update and finalize all plans.

*Initiate Applicable Staffing / Resources*

- Contact, where applicable, any functional staff to ensure that intended personnel are available and provide a "start date" for each person. If no functional organization exists, develop and implement a recruiting plan.
- Document staffing issues that may arise, coordinate their resolution and affect designated resolutions.
- Capture and track to closure all identified issues.
- Where applicable, include any critical, unresolved staffing issues in the project's risk identification/mitigation program.
- Develop and conduct training on the project requirements and plans.

*Finalize Subcontractor(s) Relationship(s)*

- Include into all subcontractor contractual documents all applicable (i.e., to the subcontractor) results of negotiations to your contract.
- Review your contract and the intended subcontract vehicles to assure that all applicable and appropriate "flow downs" have been affected; i.e., update where necessary.
- Have all subcontract changes approved by management and Contracts personnel prior to actual implementation. Ensure that the SOW and Ts&Cs ("slang" for terms and conditions) for the subcontract are in agreement with the needed requirements of the contract.
- Cause the negotiation and contract "signing" process between you and the subcontractor to take place. Ensure that the appropriate management personnel approve the intended negotiations prior to actual implementation.
- Communicate all applicable subcontracted SOW and Ts&Cs items to the project personnel responsible for tracking the subcontractor performance.
- Ensure that the results of the "definitization" of the subcontract are properly reflected into the project's planning as a whole.
- Ensure that all affected parties on the project are provided with the updated, if applicable, planning for the project.
- Ensure prime contractor IT and Security requirements are flowed down to the subcontract.

*Institute Project Progress Tracking Mechanics*

- Solidify the actual progress metrics to be collected, analyzed and used for process enhancements; assure that all your organization's requirements as well as those specified by the contract are included.
- Document all progress metrics and communicate to all project personnel.
- Ensure that the defined collection/feedback mechanics are viable; i.e., will the mechanics actually work?
- Communicate such mechanics to all project personnel.

*Establish Communications*

- Set up/solidify the means by which all project personnel may be informed of all aspects of the project's implementation affecting them in a timely manner.
- Allow for the identification and resolution visibility for any/all issues that project personnel may detect.
- Verify that the means defined will actually work.
- Explain the project communications approach/methods to all project personnel.

# Appendix C: Sample PM checklist

| No. | Task | Date |
|---|---|---|
| 1 | PM is identified. | |
| 2 | PM receives a Project Orientation, which includes the following:<br>   • Present state of the project<br>   • Primary points of contact (customer & boss?)<br>   • Technical & personnel aspects<br>   • Other pertinent information that could affect the job | |
| 3 | PM attends any necessary training: (if applicable)<br>   • PMP Training<br>   • Management Training<br>   • Accounting Training (Forecasting, estimating, profit calculation)<br>   • Any other pertinent training that could affect the job | |
| 4 | Create Configuration Management Repository (CM) to house all deliverables. | |
| 5 | Create staffing profiles | |
| 6 | Initialize Financial System labor accounts and update (from accounting) | |
| 7 | Request JSR and copy of Invoices (from accounting) | |
| 8 | Update or create Integrated Master Schedule (IMS) in the CM repository | |
| 9 | Update or create PgM in the CM repository | |
| 10 | Derive the project overall milestone schedule (high level schedule) | |
| 11 | View and update the project information:<br>   • Contract information      • System interfaces<br>   • COTS software           • Previous Audit Reports (if any)<br>   • Databases (which databases?)    • Labor accounting reports<br>   • Effort & costs (time accounting) • Software Development Plan<br>   • Critical computer resources    • Lessons Learned<br>   • Program risk areas         • Performance indicators (metrics)<br>   • Critical dependencies & tasks<br>   • Manpower & personnel<br>   • Program documentation<br>     Software size | |
| 12 | Update Authority Matrix, if needed | |
| 13 | Estimate number of hours required for each task | |
| 14 | Generate cost estimates based on resource rates and fixed costs | |
| 15 | Assign Work Breakdown Structure (WBS) functions to detailed schedule | |
| 16 | Allocate resources to each task | |
| 17 | Identify any additional metrics to track | |
| 18 | Conduct Internal Project Kickoff Meeting | |
| 19 | Conduct Customer Project Kickoff Meeting | |
| 20 | Conduct Project Management Review (PMR) as required and create minutes or close out action items | |

| No. | Task | Date |
|---|---|---|
| 21 | Inform the project personnel of assigned tasks | |
| 22 | Document internal communication activities (status reports, meeting minutes, readiness reviews, management reviews) | |
| 23 | Document external coordination activities (clearances, approval, authorizations, etc.) | |
| 24 | Take appropriate action to resolve any conflicts internal & external to the project | |
| 25 | If PM identifies change<br> • Analyze change to determine impact to project<br> • Adjust affected areas caused by change:<br>  ○ Workloads<br>  ○ Task assignments<br>  ○ Effort and estimate projections<br>  ○ Any other affected area | |
| 26 | Document any decisions that affect project progress, plan, or future events (CCB) | |
| 27 | Review detailed plan for status | |
| 28 | Document completion dates of assigned tasks in project schedule | |
| 29 | Update metrics | |
| 30 | Review metrics; compare actual vs. planned | |
| 31 | Document any decisions that affect project progress, plan, or future events (CCB) | |
| 32 | Compare project status of Actuals vs. Planned<br> • Schedule Effort<br> • Cost<br> • Other metric indicators | |
| 33 | Evaluate the performance of the project: identify which tasks were completed ahead of schedule, behind schedule, or not being met and any other lessons learned | |
| 34 | Reward individuals on the project | |
| 35 | Take corrective actions when and where applicable<br> • Explore and understand all factors involved in the problem | |
| 36 | Document any changes that affect project progress, plan, or future events (CCB) | |
| 37 | PM must generate historical record of completed cycle<br> • Historical record must be stored in CM Repository | |

# Appendix D: Automated project management tools

*Project management*

Project management software helps PMs and project teams manage and collaborate to meet project objectives while managing resources and cost. It's also used by Program Managers to oversee multiple complex projects. Your organization may already have a mandated project management tool like Microsoft Project or even a unique internal system. Depending on the complexity of your project, you'll want most or all of the following capabilities:

- Planning
- Scheduling including graphical displays
- Cost estimating
- Forecasting (based on data from previous similar projects)
- Resource allocation
- Project tracking
- Time tracking
- Task management
- Risk management
- Permission settings
- Budget and expense tracking
- Project status analysis and reports
- Collaboration
- Document sharing
- Internal messaging

One last thought...be careful to avoid tools that require excessive resources to input the data. Sometimes the pressure to use more complex tools comes from above and a wise PM will insist on the resources necessary to "feed the beast" especially if they aren't provided for in the contract!

*Financial system*

Your organization's financial and accounting system will provide Project Status Reports and other customized project-oriented financial reports. Your Financial System login credentials may be limited to designated administrators in the business units who are authorized to access and run the Financial System reports. There is a set of off-the-shelf contract financial systems in use by many contractors. However, each organization usually modifies the system fields and processes to fit their unique needs.

Project financials dashboard reports, if your organization uses them, provide access to both summary and detail project information for current and prior periods as well as valued funding data. The Dashboard design usually incorporates a summary view with drill down capabilities and filtering. Views are customized based on a user's Security provisioning.

# Appendix E: Project close-out checklist

- Verify that all contractual terms have been met.
- Verify completion and close-out of all subcontracts.
- Obtain cost information from finance and accounting.
- Conduct an inventory of the project's capital equipment.
- Transfer customer facilities and equipment back to customer control.
- Transfer staff members to other projects.
- Validate actual cost, schedule, risk, engineering, and quality metrics.
- Determine whether all project deliverables are complete, accurate and have been delivered.
- Obtain agreement from customer, corporate management, and team that project is complete.
- Assign staff to complete any remaining action items.
- Verify that action items are completed correctly.
- Verify that all project records are complete and have been archived.
- Capture and document for future reuse, as appropriate, any best practices developed and any lessons learned.
- Close all Work Breakdown Structure (WBS) numbers relating to this project.
- Validate all Security closeout actions have been completed and customer informed.

# Appendix F: Additional PM problems to avoid

*Contract Terms*

- Don't accept changes from the customer that extend the work without a contract change. If necessary to keep the peace, explain that your hands are tied!
- Don't rely on verbal (or even written) agreements between technical counterparts for out-of-scope work. These are not contractually binding. Contracts personnel must document all agreements. I learned this the hard way early in my career to the tune of over $500k in additional non-reimbursed work.

*Project Control*

- Don't change the PgM Plan without verifying that it's compatible with the contract.
- Don't let lack of performance according to plan go by for more than two weeks without investigating and taking remedial action. Things don't just straighten themselves out.
- Don't be afraid to notify senior management if your project needs resources or appears to be heading for serious trouble. And don't be defensive. Management should be there to help. Regardless, the problem(s) you hide will almost always come back to haunt you!

*Role as Manager*

- Don't let the team lose focus. Keep the focus on project objectives and priorities.
- Don't be passive in managing the schedule. Schedule slack is precious; don't give it up without a fight.
- Don't forget to seek expert help from Contracts, Legal, Pricing, Quality Assurance and Human Resources.
- Avoid calling review meetings with large groups and then only dealing with individuals on a one-to-one basis during the meeting. It will bore the others and wastes money. Also, if a group meeting is necessary, don't "dominate" it.

# Appendix G: Common project issues

*Project initiation and planning*

- Weak management approach or an unsound approach for the project proposal through start up
- Initial and follow-up planning insufficient to provide clarity and completeness
- CM/CC methods/controls insufficient to meet evolving system
- Poor project start-up definition
- Viable processes proposed and available but seldom used
- Policy adherence given "slow-roll" by all concerned
- Little up-front attention paid to necessary architecture of the intended system and, when attention is paid, no viable analyses of "workability/do ability" of the architecture defined
- Requirements definition(s) not clearly nor completely specified sufficient for the viable use of the intended implementers
- System's performance characteristics not sufficiently identified, allocated, tracked, nor well understood
- No viable/approved "business case/plan" defined or used
- Task interdependencies not defined/maintained and/or used as a management "tool"
- "Completion criteria" not defined for individual task assignments
- The PM has not fully read the contract
- Relationships with subcontractors, customer-furnished equipment, or associate contractors not included in the schedule
- The Bill of Material is inadequately defined or controlled


*Project monitoring and control*

- The PM doesn't develop nor implement a communications plan
- PM doesn't set and communicate priorities between cost, schedule and superior technical performance
- A different team is performing the program than the one that bid it and the technical approach is now deviating from the original vision
- PM doesn't believe in assigning formal action items at staff meetings
- PM doesn't effectively delegate and becomes a single-point failure, delaying all decisions

- Program team not committed to cost/schedule objectives because they view superior technical performance as more important
- Sufficient QA methods not employed
- Management reviews seldom held
- Cost based on bid productivity estimates but productivity never measured during life of project
- Metrics not tracked
- During reviews, "accomplishments" presented without being tied to things that were "planned"
- Risks and risk mitigation actions are seldom addressed at reviews
- Acceptance criteria for deliverables seldom defined/agreed upon with the customer prior to deliveries or customer acceptance of criteria is not documented
- Planning and progress tracking mostly by subjective means vs. objective measures
- Efforts in progress not consistent with baseline planning
- Capabilities continually moved to later releases to "ease" schedule pressures

*Staff management*

- PM lacking in prior, successful experience; i.e., never completed a successful PM job before or, even worse, never managed a fixed-price program
- Adequate training/mentoring of management staff not done
- Employees leaving early to seek "better opportunities"

*Risk management*

- Risks not identified early-on and/or no risk identification-mitigation program in effect
- Senior management doesn't pay attention to project risks...only reacts to crisis's. Basically, praying things are okay or will get better!

# Appendix H: Business partnering lessons learned

As a PM, you'll often either lead or support the identification of business partners as a part of your project execution or new business development capture activities. When seeking teaming partners or solution vendors, take advantage of lessons learned from past teaming experiences. The following paragraphs provide a brief summary of the most significant teaming lessons learned I've gathered and can help you avoid some pretty costly mistakes!

*Small business teaming lessons*

- Small companies, unlike larger ones, don't always have the cash flow needed to support their payroll and accounts payable. Be prepared to work with finance and accounting to ensure your organization's timely payment of small business subcontract invoices.

- Small businesses are sometimes "one deep" in experts. Unless they are highly focused, most small businesses will not have the depth of resources needed if your project gets in trouble. The best approach is to provide early communication of expanding project needs to the subcontractor. Hopefully, they'll be able to staff up (or identify qualified candidates) in anticipation of the need.

- Small business support costs vary. Expect general-purpose small businesses to offer significantly lower pricing with varying degrees of expertise. Expect highly focused small businesses (Security consultants, SAP specialists, software analysts, etc.) to be significantly higher priced than your comparable internal resources.

- Be prepared to mentor Small, Women-Owned, Veteran-Owned and other Disadvantaged Businesses. These businesses are often start-ups. They need (compassionate as opposed to arrogant) coaching and mentoring from your organization to help them understand how the subcontracting, payment, invoicing, payment and related processes work.

*Potential competitor/large business teaming lessons*

- A large company can make an excellent partner, but you need to get their attention and convince them you can win a competition. The best way to do this is to have a well-defined value proposition in place that indicates why you can win, what's in it for them, what's in it for you and the follow-on business potential. If it's a win-win scenario, the partner is more likely to provide good support.

- Ensure a balanced sharing of risks and rewards. Some large subcontractors will want to make your organization take on all of the expense and risk for bid and proposal and then take advantage of any contract that results. If the large company is unwilling to share the investment burden as it relates to their share of the resulting work, then get them to reduce their prices or negotiate them down to a lower work percentage.

- Regulate access to the customer. If your organization is the prime contractor, you are ultimately responsible for the project's outcome. As the PM, you must ensure that subcontractor personnel perform in a fashion supportive of project success and controlled by your organization. If the customer starts to provide tasking directly to the subcontractor, your organization may be unable to deliver on its contractual promises. Document all subcontractor interactions with the customer and make sure the customer goes through your organization for any change requests, issue resolutions, or problem discussions.

*Product selection lessons*

- Beware of beta products! Ensure that your organization has performed an in-house, hands-on evaluation of a beta product and that it meets its performance specifications prior to proposing it to a customer. A product that is vaporware usually lacks documentation or lacks vendor technical support could seriously compromise your project performance.

- Understand the difference between the best technical product and a "market leader" product. Your organization may lean toward the best technical product due to your staff's technology orientation. This focus may result in a product selection contrary to the customer's risk tolerance.

- The customer will want a product from a vendor that will be around to support the product for years to come; a company that doesn't sell in high volumes or build market share might not survive.

- The customer will want to be sure they can find personnel with the product skills needed to maintain and enhance the product; if the product isn't widely utilized, these personnel resources may be both scarce and expensive, creating a supportability risk situation for the customer's organization.

- The customer will want to keep pace with technology change; a vendor with low revenues won't be able to afford to invest in major feature additions and upgrades.

- The customer wants timely product technical support; a smaller vendor with fewer resources might not be able to respond to customer needs. Think customer satisfaction first, and technology second.

- The good news is that your organization probably has lots of bright technologists. The bad news is that they can't keep up with every new product that hits the marketplace. This means your organization will be dependent on the vendor during your initial engagement using their product.

140

As a result, look for vendors that can supply training, training materials, documentation, sample products and technical consulting support during your first project exposure. This will minimize your organization's risk of non-delivery and cost overruns.

- Beware of Start-ups! This is a corollary to the above lessons. In the Internet economy, start-ups are everywhere. Take advantage of any technology market analyst services subscriptions to verify likely winners. Investors and venture capitalists can also be a good source of information.

- Many product vendors will give your organization better product pricing if you can arrange it so the order is placed near the end of the month or end of the quarter. The vendor sales representative wants to make a good showing for the period. If you can offer the order now, instead of waiting until the next month or next quarter, they'll be happy to negotiate with you. Work with procurement to get the best deal for you and the customer.

- Take advantage of quantity buys. Investigate purchasing a large quantity of product up front rather than smaller quantities spaced over several months or quarters. As above, you may find that you can get a much better price from a total buy perspective.

- Leverage your organization's buying power. Find out what other products your organization has bought from the vendor. Use this information to negotiate significant product discounts for a new purchase or take advantage of volume pricing agreements already in place.

# Appendix I: PM new business pursuit guidelines

The guidance in this appendix directly contributes to: winning new business, defending your existing business, and achieving greater customer intimacy. As a PM, you're the most important link in this chain due to your proximity and continual contact with your customers.

*Business opportunity support – "do's"*

- Assist (through performance) the customer in valuing your organization's services and solutions. They will then come to you if they have a need. Bottom line...do a great job.

- Forward information (if you're at liberty) to BD or your Manager concerning potential opportunities that are coming up in the future.

- Forward information to BD or your Manager concerning innovative business ideas and/or concepts that you have developed and that you see having a potential place with your customer or other customers (these will also be used in the re-compete process).

- Assist your manager in getting the information together for potential case studies, past performance and the identification of Subject Matter Experts (SMEs) who will assist on future proposal efforts.

*Business opportunity support – "don'ts"*

- Don't directly "work" your customers in a sales capacity (don't schmooze them).

- Don't forward information immediately in advance of a scheduled release (or after a release) if you're in a position that precludes your open communication with your organization, i.e. "trusted agent" status/walled off from your organization due to you having an acquisition sensitive position on a procurement.

- Don't sacrifice project performance in pursuit of new business. On the other hand, most organizations expect PMs to enthusiastically support the pursuit of business. It's an ongoing balancing act that requires patience and finesse!

# Appendix J: Glossary of terms

**Account** – An account is directly in support of a contract. Requires a Project ID, Activity ID, Res. Type, Res Category that is NOT a direct expense.

**Activity ID** - Each activity ID is identified with a project #. You can have many activity IDs for a project. Accounts may be further broken down into individual IDs for specific labor categories, indirect and direct charges.

**Activity** - A component of work performed during the course of a project.

**Assumptions** - Factors that, for planning purposes, are considered to be true, real, or certain without proof or demonstration.

**Award Fee** - Profit earned on an award fee type of contract with the amount being determined by the customer as a percentage of a fixed award fee pool. The percentage is established based on the contractor's performance against a set of award fee criteria, which may be quantitative in nature but most often are subjective.

**Baseline** - An approved plan for a project, plus or minus approved changes. It's compared to actual performance to determine if performance is within acceptable variance thresholds. Generally refers to the current baseline but may refer to the original or some other baseline.

**Basis of Estimate (BOE)** - The rationale used to determine the cost of a particular task or item in a proposal. Often BOEs must be provided with the proposal.

**Budget at Completion** - The total amount of money allocated to do all authorized work.

**Burden** – Each burden ties back to a Group, Operating Unit and Department overhead. For instance, your organization's indirect facility costs, government site indirect costs and G&A costs.

**Change Control** - Identifying, documenting, approving or rejecting, and controlling changes to the project baselines.

**Confidentiality Agreement** - See Non-disclosure agreement.

**Contingency Reserve** - The amount of funds, resources, or time needed above the estimate to reduce the risk of overruns of project objectives to a level acceptable to your organization.

**Cooperation Agreement** - Similar to a teaming agreement but used mostly for pursuing a broader business objective. Often used as a framework for setting up a series of teaming agreements for different purposes but within the scope of the cooperation agreement.

**Cost Control Account** - A collection of work packages that are commonly managed and identified in the project WBS for that purpose.

**Cost/Cost of Sales** - The total of all costs incurred in producing or acquiring products or services for sale. The sum of direct and indirect costs, as seen from your organization's perspective, for a product or service provided to a customer (i.e., how much it costs your organization to perform on a project).

**Cost Performance Index** - The ratio of earned value (budgeted cost of work performed or BCWP) and actual costs (actual cost of work performed or ACWP). CPI = BCWP/ACWP.

**Dept. ID** Each Operating Unit has its own specific Dept. IDs. The system will only take Dept. IDs for that Op Unit.

**Estimate at Complete** - An estimate of the cost at completion of the contract given the schedule and current progress (also called Latest Revised Estimate or LRE).

**Fixed Fee** - A predetermined, fixed amount of profit such as on a cost plus fixed fee contract. The fee isn't a function of the actual cost but may be proposed and negotiated as a percentage of negotiated baseline cost.

**Government Wide Acquisition Contract (GWAC)** – a contract vehicle that more than one government agency can use to procure services and products.

**Gross Margin** - The difference between net sales and cost of sales

**Gross Margin Percentage** - Ratio of gross margin divided by net sales expressed as a percentage.

**Historically Black Colleges and Universities** - An institution that has been determined by the Secretary of Education to meet the requirements of 34 CFR 608.2. The term also means any nonprofit research institution that was an integral part of such a college or university before November 14, 1986 (DFAR 252.219-7003).

**Indirect Costs** - Costs for conducting and sustaining business that are collected at above the project and allocated to projects within your organization. This includes the associated proportional costs for administrative support, information systems, marketing/business development investment, facilities, support infrastructure, and senior management. This is typically expressed as a percentage of direct cost to facilitate computation of the total project cost, or total project cost of sales.

**Latest Revised Estimate** - an estimate of the cost at complete given the schedule and current progress (also known as Estimate at Complete or EAC).

**Limitation of Funds** - A contract clause used to limit the amount of funds allocated to a contract. This clause may require the contractor to notify the customer at a specified amount of time before authorized funds are completely expended.

**Limitation of Liability** - A contract clause that limits the contractor's liability.

**Mentor-Protégé Program** - DoD program that provides incentives to major DoD contractors to assist small disadvantaged business (SDB) firms in enhancing their capabilities of performing as prime and subcontractors under DoD contracts. The program seeks to increase the participation of SDBs and subcontractors under DoD and other federal government and commercial contracts. The program encourages SDBs, and DoD contractors to establish long-term business relationships free from government interference and existing government restrictions on the relationships between large and small businesses.

**Milestone** - A milestone is a key event of the project specified in the Request for Proposal (RFP), the SOW, and/or the response to the RFP. Examples of milestones types are:
- A significant point in systems development or systems integration
- An accomplishment that must be attained in order to meet stated objectives
- An event that singularly defines where you're – verifiable by others
- Any event that requires approval before proceeding further
- A decision point that defines the course of further activity in the project
- Any event requiring verification of interface compatibility outside the implementers' group

**Minority Institutions** - Accredited colleges, universities or enrollment of a single minority group or a combination of minority groups (underrepresented in science and engineering) exceeds 51% percent of the total enrollment. (DFAR 252.219-7003)

**Monitor** - Collect project performance data with respect to a plan, produce performance measures, and report and disseminate performance information.

**Net Profit** - The difference between periodic revenues and matching costs (including all taxes and interest paid).

**Net Sales** - Total revenue from sales for a specified period, less adjustments such as returns, allowances, and sales discounts.

**Non-disclosure Agreement** - A documented agreement between two parties (companies) containing rules and regulations for handling each other's proprietary information.

**Performance Management Baseline** - The collection of work scope, cost and schedule against which progress will be measured.

**Point of Total Assumption** - On a fixed price incentive contract, the cost at (and above) which the contractor assumes the responsibility for the added cost. In effect, the contract converts to a firm fixed price contract in an overrun condition. All costs above the PTA are a profit loss.

**Portfolio** - A collection of projects or programs and other work that are grouped together to facilitate effective management of that work to meet strategic business objectives.

**Portfolio Management** - The centralized management of one or more portfolios, which includes identifying, prioritizing, authorizing, managing, and controlling projects, programs, and other related work, to achieve specific strategic business objectives.

**Price** - The transactional value of a project as seen from the customer's perspective (i.e., what your organization charges the customer for the project).

**Probability and Impact Matrix** - A common way to determine whether a risk is considered low, medium, or high by combining the two dimensions of a risk – probability of occurrence and impact on objectives if it occurs.

**Profit** - The difference between periodic revenues and cost of sales.

**Project ID** - Each project ID is identified with a specific customer contract. You can have many projects to one contract, but the contracts tie back to the legal entity that was awarded the contract.

**Program** - A group of related projects managed in a coordinated way to obtain benefits and control not available from managing them individually.

**Program Management** - The centralized coordinated management of a program to achieve the program's strategic objectives.

**Project** - A temporary endeavor or task undertaken to create a unique product, service, or other result.

**Project Management** - The application of knowledge, skills, tools, and techniques to project activities to meet the project requirements.

**Project Management Office (PMO)** - The organization body or entity assigned various responsibilities related to the centralized and coordinated management of those projects under its domain. The responsibilities of a PMO can range from providing project management support functions to actually being responsible for the direct management of a project or group of projects.

**PgM Plan** - A formal, approved document that defines how the project is executed, monitored, and controlled.

**Project Risk Management** - All processes concerned with planning, identifying, analyzing, responding to, monitoring, and controlling project risks.

**PRO-Net** - A database administered by the Small Business Administration. The database that contains supplier profiles of small, small disadvantaged and women-owned firms.

**Residual Risk** - A risk that remains after risk mitigation measures have been implemented.

**Revenue** - The recorded incidence of a sale of products and/or services as recognized by your organization's accounting system (i.e., the money paid by a customer to your organization for performance on a project).

**Risk** - An uncertain event or condition that, if it occurs, has a negative effect on a project's objectives.

**Risk Management Plan** - The formal document describing how project risk management will be structured and performed on the project. It's contained in or is a subsidiary plan of the project management plan.

**Risk Tolerance** - The degree, amount, or volume of risk that your organization or individual is willing to withstand.

**Schedule Performance Index** - The ratio of earned value (budgeted cost of work performed or BCWP) and budgeted cost of work scheduled (BCWS). SPI = BCWP/BCWS.

**Small Disadvantaged Business Concern** - A small business that is at least 51% owned by an individual who is both socially and economically disadvantaged, as defined by the Small Business Administration (13 CFR part 124), the majority of earnings which directly accrue to such individuals. An SDBC can be publicly owned, with at least 51% of the stock unconditionally owned by one or more disadvantaged individuals. One or more such individuals must control the daily business operations. Disadvantaged individuals include African-Americans, Hispanic Americans, Native Americans, Asian-Pacific Americans, Subcontinent Asian Americans (FAR 52.219).

**Technical Performance Measure** - A metric for assessing how well a technical requirement is being met.

**Termination Liability** - A specified dollar amount usually spread over the time to cover a customer's liability if a contract is cancelled prematurely. Often the contractor is required to notify the customer at a specified amount of time before authorized funds are completely expended. The contractor should take into account the amount identified for termination liability in this notification.

**Woman-Owned Small Business** - A small business that is at least 51% owned by one or more women and whose management and daily business operations are controlled by one or more women. (FAR 52.219).

# Appendix K: Acronyms

**ACWP** - Actual Cost of Work Performed

**BAC** - Budget at Completion

**BOE** - Basis of Estimate

**BCWP** - Budgeted Cost of Work Performed

**BCWS** - Budgeted Cost of Work Scheduled

**BOA** - Basic Ordering Agreements

**BOE** - Basis of Estimate

**BOM** - Bill of Material

**BAC** - Budget at Completion

**CCB** - Configuration Control Board (or Change Control Board)

**CDR** - Critical Design Review

**CDRL** - Contract Data Requirements List

**CFE** - Customer Furnished Equipment

**CFI** - Customer Furnished Information or Customer Furnished Items

**CM** - Configuration Management

**DM** - Data Management

**CMS** - Contract management System

**COTR** - Contracting Officer's Technical Representative

**CPAF** - Cost Plus Award Fee type of contract

**CPFF** - Cost Plus Fixed Fee type of contract

**CPIF** - Cost Plus Incentive Fee type of contract

**CV** - Cost Variance

**EAC** - Estimate at Complete

**ETC** - Estimate to Complete

**EVMS** - Earned Value Management System

**FAR** - FAR

**FFP** - Firm Fixed Price type of contract

**FPI** - Fixed Price Incentive

**G&A** - General and Administrative

**GFE** - Government Furnished Equipment

**GFI** - Government Furnished Information

**GWAC** – Government Wide Acquisition Contract

**HBCU** - Historically Black Colleges and Universities

**HW** - Hardware

**IDIQ** - Indefinite Delivery, Indefinite Quantity

**IFR** - Internal Funds Request

**IPPT** - Integrated Product Process Team

**ITAR** - International Traffic in Arms Regulations

**LOE** - Level of Effort

**LRE** - Latest Revised Estimate

**NDA** - Non-disclosure agreement

**ODC** - Other Direct Costs

**PCA** – Project Control Analyst

**PgM** – Program Management

**PM** – PM

**PMB** - Project Management Baseline

**PMBOK®** - Project Management Body of Knowledge

**PMCoP** - Program Management Community of Practice

**PMR** - Program Management Review

**PR** - Purchase Requests

**PRA** - Project Review Authority

**PTA** - Point of Total Assumption

**PWA** - Project Work Authorization

**QA** - Quality Assurance

**QC** - Quality Control

**SB** - Small Business

**SDB** - Small Disadvantaged Business

**SDR** - System Design Review

**SDRL** - Subcontract Data Requirements List

**SPI** - Schedule Performance Index

**SOW** – Statement of Work

**SV** - Schedule Variance

**SW** - Software

**SWOT** - Strengths, weaknesses, opportunities, and threats

**T&M** - Time and Materials

**TPM** - Technical Performance Measure

**Ts&Cs** - Terms and Conditions

**VOB** - Veteran-Owned Business

**WOB** - Woman-Owned Business

**WBS** - Work Breakdown Structure